The Secrets of

Great
Botanists

First published 2018
This edition published 2024
Exisle Publishing Pty Ltd
PO Box 864, Chatswood, NSW 2057, Australia
226 High Street, Dunedin 9016, New Zealand
www.exislepublishing.com

ISBN 978-1-923011-03-8

Conceived, designed and produced by
The Bright Press, an imprint of the Quarto Group
1 Triptych Place
London SE1 9SH
United Kingdom
www.quarto.com

Publisher: James Evans
Editorial Director: Isheeta Mustafi
Managing Editor: Jacqui Sayers
Art Director: James Lawrence
Project Editor: Jemima Solley

Design and image research: Lindsey Johns

Printed in Malaysia

2 4 6 8 10 9 7 5 3 1

**This book is dedicated to two great plantsmen, Roy Lancaster CBE,
VMH, FCIHort, FLS and the late Mark Flanagan DHE (Hons.),
M. Hort. (RHS), MVO, VMH, for so generously sharing their knowledge
and inspiring everyone who loves plants.**

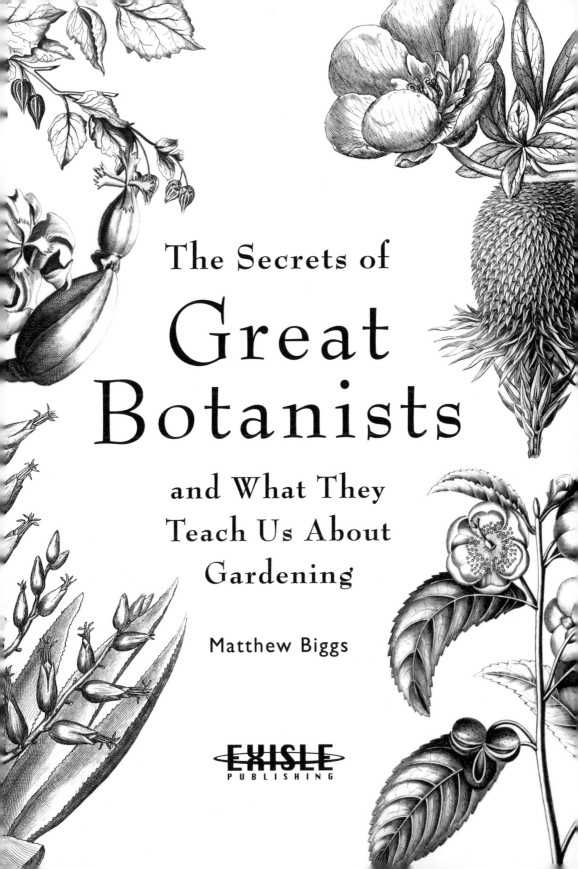

The Secrets of
Great
Botanists

and What They Teach Us About Gardening

Matthew Biggs

EXISLE
PUBLISHING

Contents

Colchicum autumnale
(meadow saffron, see p22)

Meconopsis betonicifolia
(blue poppy, see p166)

Introduction

Welcome to the lives of 36 great botanists, past and present, whose study of plants revolutionised the scientific world and increased our understanding of the importance of plants for our survival. All of the botanists included in this book were passionate about their particular specialism within this vast subject. Whether they were searching for plants in wild places, prescribing herbal medicines or forming new structures of plant classification, they were driven by an insatiable desire to learn and discover, so that humans might benefit from improved crops, medicines or gardens. When eminent naturalist Sir David Attenborough reopened the restored Temperate House at the Royal Botanic Gardens, Kew, on 3 May 2018, he remarked, 'the best gardener is a good botanist and the best botanist is a good gardener'. It is necessary to have an understanding of both disciplines, if we are to have a full understanding of plants.

BOTANY FOR CHANGE

Much is written about gardening and the aesthetic use of plants, but less is said about the scientific study of botany or those botanists who changed the world of gardening and beyond. Botanists are not just white-coated scientists who spend their lives in laboratories, squinting down microscopes at the cells of plants – they are the explorers, the trailblazers and the inventors, advancing a range of fields through their studies of the natural world.

Among the huge number of iconic botanists who have shaped the way we live today are those who studied plant-based medicines; cytologists, who researched cells and chromosomes; plant breeders and geneticists, who developed economic crops; field botanists, who endured extreme hardship while collecting species that were new to science; palaeobotanists, who studied fossils, and ethnobotanists, who lived and learned among indigenous peoples and their plants.

Dr Richard Evans Schultes (see pp194–199), a pioneering ethnobotanist, lived and worked among the plants and people of the Amazon rainforest.

A HISTORY OF BOTANY

For the majority of our history as a species, we lived as hunter-gatherers. Selecting useful plants would have been a process of trial and error; they would have eaten then rejected some plants that were unpalatable, even deadly, and continued to acquire other plants that were tasty or provided other benefits. In time, people settled where food was plentiful and discovered that seeds of the best plants could be collected, cultivated and selected to safeguard supplies and ensure their survival. It was only once civilisations and empires were established that widespread scientific research began. Early plant studies began with medical botany. Physicians needed to identify and record the characteristics and doses of plants as treatments; when the results were documented, the first 'herbal' books appeared, describing the medicinal properties of herbs. The Greeks began the tradition of questioning and deduction that led to scientific ways of thinking. Theophrastus (c. 372–c. 287BC), a great philosopher, wrote over 200 works, including two on botany. His *Enquiry into Plants* dealt with the description and classification of about 550 plant species and his *On the Causes of Plant Phenomena* discussed the physiology and reproduction of plants. Both books were influential from the time of Dioscorides (c. AD 40–90, see pp8–13) to the Renaissance.

THE FINAL CUT

So, which botanists were chosen for this book? It was inevitably a personal selection, made from hundreds of iconic botanists from around the world, from ancient times to the present day. Although the choices were subjective, it was important to represent a wide range of disciplines and contributions to botany. Instead of providing 'hands-on' lessons for gardeners, this book aims to teach readers about the observations and ideas of groundbreaking botanists, and show how influential they have been on gardening and the wider world. The 'Inspiration for Gardeners' pages at the end of each botanist's section detail plants and discoveries related to the botanist, to inspire readers to try these plants and tips at home.

De Materia Medica (c. AD 30) by Dioscorides (see pp8–13), was translated into many languages, including Arabic. It became the basis of many medieval books on botany.

Pedanius Dioscorides

DATE	C. AD 40–90
ORIGIN	PRESENT-DAY TURKEY
MAJOR ACHIEVEMENT	*DE MATERIA MEDICA*

Dioscorides, the 'Father of Pharmacology', was a herbalist with the Roman Army, and one of the first to observe and record plants in detail and to recognise the uses of medicines from all three kingdoms of the natural world – animal, vegetable and mineral. His *De Materia Medica* ('On Medical Materials') was laid out by category and then by the physiological effects they had on the body, a format that he believed would make it easy to learn and retain the information. This was proven in practice and it became the basis of medicine for 16 centuries.

Euphorbia characias,
Mediterranean spurge

Pedanius Dioscorides was born in Anazarbus, a small city northeast of Tarsus in the Roman province of Cilicia (now Turkey), at the time of Nero and Vespasian. His travels as a surgeon with the Roman Army into Italy, Greece, northern Africa, Gaul, Persia, Egypt and Armenia provided him with a wealth of opportunities to study the features, distribution and medicinal properties of plants, animals and minerals.

Dioscorides started *De Materia Medica*, a list of known medicinal plants of the Roman Empire and new introductions, around AD 50, finishing it roughly 20 years later. He would collect samples of local medicinal or useful herbs as he travelled, many of them previously unknown to Greek and Roman physicians, describing their restorative effects and botany, including roots, foliage and sometimes flowers. Dioscorides was also one of the first writers to observe and identify plants at all stages of their growth, and in all seasons, noting variations due to climate, altitude and precipitation. In addition, he offered practical advice on collecting and storing drugs, as well as their preparation and dosage; he even trialled drugs clinically, and thoroughly researched traditional usage of herbs in the locations where he found them. His writings included information gleaned from oral traditions and drawn from earlier texts, including 130 plants from the Hippocratic Collection, and others from the work of the ancient Greek physician Crataeus.

Dioscorides' original non-illustrated manuscripts, which no longer exist, contained medical information on around 600 plants, 90 minerals and 35 animal products (including two using viper's flesh; one being pickled in wine, oil, dill and salt for calming nerves and sharpening eyesight). These were divided into five volumes. Book One covered aromatics, oils, salves, trees and shrubs; Book Two dealt with animals, animal parts and products, cereals and herbs; Book Three discussed roots, juices, herbs and seeds; Book Four addressed additional roots and herbs; and Book Five housed information about wines and minerals.

BODILY AILMENTS

As you would expect, painkillers feature prominently in this work. For example, Dioscorides wrote of the willow (the *Salix* species, a natural remedy for pain and fever, and the basis of aspirin): 'a decoction of them is an excellent fomentation for gout'. He also alerted readers to the difference between a therapeutic and toxic dose of that powerful painkiller and dangerous narcotic, opium: 'a little of it, taken as much as a grain of ervum [probably the seed of *Vicia ervilia*, known as ervil or bitter vetch, an ancient legume found in the Mediterranean region], is a pain-easer, and a sleep-causer, and a digester ... but being drank too much it hurts, making men lethargicall, and it kills.'

There are also several remedies for toothache, including the resin of *Commiphora* (myrrh); the bark of *Platanus* (plane trees) soaked in vinegar; a decoction of tamarisk leaves mixed with wine; oak galls; the resin of *Rhus* (sumac); a decoction of mulberry leaves and bark; latex taken from the fig tree and *Euphorbia characias*

(Mediterranean spurge, pictured on p8) mixed with oil; and a decoction of asparagus roots and *Plantago* (plantain). If the problem could only be solved by extraction, olive oil sediment mixed with the juice from unripe grapes and cooked to the consistency of honey was to be smeared on decayed teeth to loosen them.

Dioscorides' manuscript reveals much about other uses of plants in the Roman Empire, too – for example, young *Pistacia lentiscus* twigs used for dental hygiene, shampoo made by pounding henna leaves soaked in the juice of soapwort (which acted as a foaming agent) and soot from the burnt resin of conifers used to darken eyebrows and eyelashes. And he often warns against adulteration, mentioning methods and the means of detection: frankincense, which was frequently adulterated with pine resin and gum, and valerian root, which when adulterated with butcher's broom became hard, difficult to break and lost its pleasant aroma.

A LASTING HERITAGE

Remarkably, *De Materia Medica* remained the leading herbal text for 16 centuries, proving the accuracy of its author's observations and instructions through many generations. During the Middle Ages it was copied by hand, illustrated and translated into Arabic, Greek and Latin. From the sixteenth century, translations extended to Italian, German, Spanish and French, and an English version by the botanist John Goodyer appeared in 1655. Goodyer, together with his assistant John Heath, worked on his translation from 1652 to 1655, consulting at least 18 editions of *De Materia Medica* along the way. The pair worked quickly, beginning the first volume (728 pages) on 27 April 1652, and moving on to the second by 28 March 1653. *De Materia Medica* also formed the basis for later herbals written by the likes of Leonhart Fuchs, Matthaeus Lobelius, Carolus Clusius, John Gerard and William Turner, though these authors' own thoughts and observations gradually displaced Dioscorides'original text.

Salix alba (white willow). Dioscorides recommended the use of willow to treat gout.

However, in 1934 Sir Arthur Hill, Director of the Royal Botanic Gardens at Kew, found the original work still in use on Mount Athos: 'The official botanist monk ... a remarkable old man with an extensive knowledge of plants and their properties ... travelled very quickly, usually on foot, and sometimes on a mule, carrying his Flora with him in a large black bulky bag ... his Flora was nothing less than four manuscript volumes of Dioscorides, which apparently, he himself had copied out. This Flora he invariably used for determining any plant which he could not name at sight, and he could find his way in his books – and identify his plants to his own satisfaction – with remarkable rapidity.'

The oldest surviving and most famous copy is the so-called 'Vienna Dioscorides', created for Princess Juliana Anicia, daughter of the Western Roman Emperor Olybrius and now held in the Austrian National Library. This richly illustrated Byzantine version was completed in Constantinople in 512, and contains 383 paintings of Mediterranean plants, many of which can still be recognised today.

Pedanius Dioscorides:
INSPIRATION FOR GARDENERS

❖ Dioscorides recorded the use of euphorbia latex in dentistry but this sap can be harmful, causing inflammation of the skin, so it would have to be applied with care to avoid any soft tissue. The difference between a therapeutic and toxic dose can be minimal, when ingested, too, so always treat plants with respect. Do not self-medicate if you are uncertain about the dosage, or take expert advice before doing so. Familiar plants with common uses – such as mint and parsley for garnishes – are not eaten in large enough quantities to be dangerous.

BELOW Dioscorides prescribed the stem of *Physalis alkekengi* (Chinese lantern) as a sedative, and its berries as a diuretic.

❖ The sap of several plants can cause problems. Wear goggles when pruning euphorbia to protect your eyes. Rue causes blistering in the presence of sunlight, so prune or harvest on a dull day and cover your hands and arms as a precaution. When pruning ivy, wear gloves, goggles and a face mask; old plants contain dust, and the hairs on the leaves can trigger asthma attacks.

❖ Although Dioscorides prescribed the berries of *Physalis alkekengi* as a diuretic, fruiting plants should be sited with care. If you have young children, or young visitors to your garden, position plants with tempting berries away from places where

they can be reached by tiny hands. If you suspect poisoning, seek medical help immediately, taking a sample of the plant to the doctor or hospital for identification.

✤ According to Dioscorides, a poultice of fresh bay laurel leaves relieved the sting of wasps and bees, and could be drunk in wine to treat a scorpion sting. The juice of the leaves mixed with old wine and rose oil was also recommended as eardrops to improve hearing, while the bark would shatter kidney and bladder stones and heal an unhealthy liver. Modern drugs have replaced such remedies, though, and today the bay is used mostly as garden ornamentation and an ingredient in a bouquet garni.

✤ Dioscorides mentioned several Mediterranean herbs in his text, including lavender, rosemary and dill. Harvest these herbs early in the morning before the essential oils start to evaporate in the heat of the day, and put your clippings straight into a polythene bag. Mediterranean herbs should be grown 'hard', in poor, free-draining soil in an open sunny site, encouraging compact growth and a greater concentration of oils.

TOP RIGHT Widely used today, Dioscorides is believed to have been the first person to record the healing properties of *Lavandula* (lavender).

RIGHT *Laurus nobilis* (bay tree) can be medicinal, edible and ornamental. Dioscorides would also have known of its use in a victor's wreath.

Leonhart Fuchs

DATE 1501–1566	
ORIGIN BAVARIA	
MAJOR ACHIEVEMENT *DE HISTORIA STIRPIUM*	

Fuchs made his name in botany with the publication of his 1542
De Historia Stirpium ('Notable Commentaries on the History of Plants').
First published in Basel, this masterpiece of Renaissance botany is widely
recognised as one of the most beautiful books ever printed, due to its
exquisite hand-coloured illustrations. It was also the first time that
many new species from the New World – including now-familiar and
widespread plants like tomatoes, corn and chillies – were described
and illustrated in a book. Despite its beauty, only around 100 copies
are known to have survived.

Gentiana lutea,
great yellow gentian

Leonhart Fuchs, a childhood genius, was born in Bavaria in 1501 at the height of the German Renaissance. He gained a BA at the age of 14, opened his own school, then at 18 attended Ingolstadt University in Bavaria, studying classics, philosophy and medicine. He became a physician in Munich in 1524 and was later summoned to Ansbach to become the personal physician of the Margrave of Brandenburg. It was during this time that he also became established as a writer.

Fuchs first found fame by creating a cure for the fatal and fast-acting English sweating sickness, using a number of plants, including rosemary, butterbur and various types of gentian (such as *Gentiana lutea*, pictured on p14). In 1535 he became Professor of Medicine at the University of Tübingen, a post he held for the rest of his life. Teaching was Fuchs's priority: he brought many innovations to the curriculum, including botanical field trips for medical students.

Fuchs was passionate about his subject: 'there is nothing in this life pleasanter and more delightful than to wander over woods, mountains, plains, garlanded and adorned with flowerlets and plants of various sorts, and most elegant to boot, and to gaze intently on them. But it increases that pleasure and delight not a little, if there be added an acquaintance with the virtues and powers of these same plants.'

A PICTURE TELLS A THOUSAND WORDS

In the preface of *De Historia Stirpium* Fuchs explained that he was writing for the benefit of his fellow physicians, lamenting the difficulty of finding one physician in a hundred with accurate knowledge of even a few plants: 'I have compiled these commentaries on the nature of plants with the utmost care as well as expense, emulating the devotion of the eminent scholars mentioned above. In the first place, we have included whatever relates to the whole history of every plant, with all the superfluities cut out, briefly and, we hope, in the best order, one we shall follow regularly. Then, to the description of each plant we have added an illustration. These are lifelike and modelled after nature and rendered more skillfully, if I may say so, than ever before. This we have done for no other reason than "a picture expresses things more surely and fixes them more deeply in the mind than the bare words of the text".'

His book was one of the first to include images drawn from nature, presenting plants at several stages during their life cycle, showing seeds, roots, flowers and fruits. Plants were listed alphabetically according to the Greek alphabet. The illustrations were to be used for identifying plants, and the medicinal uses were taken from ancient herbalists, primarily Dioscorides, then Galen and Pliny (always in that order), sometimes with Fuchs's own observations added. He notes, for example: 'a drachma in weight' of the thistle root taken in wine, is of benefit against the contagion of pestilence or sweating sickness. Furthermore, the same 'steeped in vinegar, is helpful against scabies, impetigo, and all blemishes of the skin difficult to cure, if they are washed with the decoction. The same also helps toothache.'

The illustration of *Capsicum annuum* (ornamental pepper) in *De Historia Stirpium* includes its root system. Fuchs wanted to understand the whole plant.

A NEW WORLD OF PLANTS

Fuchs grew many of the plants featured in his herbal in his Tübingen garden, including the new and exotic. Of around 500 species, 100 were illustrated for the first time, including several from the New World, such as cacti, pumpkins, marigolds, potatoes, squashes, kidney beans, tobacco, corn and chillies, the specimens probably coming from Fuchs's garden. Fuchs also broke with tradition by being interested in illustrating plants exactly as they looked in nature, hand-painted with a watercolour wash under his supervision for the greatest possible accuracy. Sadly, his original set of drawings was lost.

Unusually, there are portraits of the three artists at the conclusion of the book: Albrecht Meyer, who drew the plants from living specimens; Heinrich Füllmaurer, who transferred the drawings into woodblocks; and Vitus Rudolph Speckle, 'by far the best engraver in Strasbourg', who cut them into wood for printing.

The book is also believed to be the first herbal to include a botanical glossary in the form of its introductory chapter, referred to as an 'Explanation of Difficult Terms' – 132 in all – but this helpful addition was only included in the Latin versions. The glossary also sheds light on many horticultural ideas of the time, like *Toparium*: 'topiary work which arranges trees, shrubs or herbs, into arches or vaults

Unusually, Fuchs's book depicts the artist-craftsmen who illustrated the work.

for decoration. Hence those trees and herbs should be called topiariae which are particularly adapted to this work owing to their natural flexibility and pliancy.' Or propagation, or *Propago*: 'an old vine bent down and buried in the earth in the form of arches, so that from one, many vines grow...'. And *Tomentum*: 'By tomentum the Latins meant anything with which mattresses could be stuffed to make them softer and warmer, whether this be wool or feathers, or anything else one wishes, suitable for making them softer and the body warm. So the leaves of *Dictamnus* which seem to be soft are called *tomentitia* and *lanea* by Dioscorides.'

Fuchs's herbal was very popular. It was abridged from Latin into a German folio edition (though still weighing about 5kg/11lb), then into field guides; in 1545 an edition containing only plates was published for the illiterate. During Fuchs's lifetime there were 39 imprints in various formats in Latin, German, French, Spanish and Dutch, and it was translated into English 20 years after his death. English artist and designer William Morris owned a copy and is said to have taken inspiration from its prints for some of his designs – 'I have a poor copy of it which I have found very useful: it is a rare book; but for refinement of drawing it is the best of all as one would expect from the date.'

Corn was illustrated and labelled with the name *Turcicum frumentum* in *De Historia Stirpium*.

Leonhart Fuchs:
INSPIRATION FOR GARDENERS

❖ Fuchs never saw a fuchsia; the plant was named in his honour more than a century after his death. The first species to be found was *Fuchsia triphylla*, on the Caribbean island of Hispaniola in around 1696, by the French Franciscan monk and botanist Charles Plumier, during his third expedition to the Greater Antilles. Plumier made detailed drawings of the flowering plants and ferns he saw and also revived the ancient practice of naming genera after people. The genus *Plumeria*, commonly known as frangipani, was in turn named after him.

❖ Fuchs often commented on where he saw plants. He said that clove pink was 'widely grown in pots' and 'you seldom find a house that does not have one displayed in front', for example. Basil was grown 'in clay pots on the windowsill' and the hart's tongue fern was 'prized in most gardens'. Fuchs also recorded the popularity of the chilli pepper – 'found almost everywhere in Germany now, planted in clay pots and earthen vessels. A few years ago it was unknown'.

TOP LEFT *Fuchsia triphylla* was the first fuchsia species discovered by the botanist Charles Plumier. Plumier named the genus after Leonhart Fuchs.

LEFT Fuchs recorded that *Asplenium scolopendrium* (hart's tongue fern) was 'prized in most gardens'. It is still a reliable choice today, in damp or dry shade.

✤ Fuchs included a description of corn, or maize, as one of the new introductions to Europe at the time. Elsewhere in the world it was a plant with a long history of cultivation as a staple crop. As Nikolai Vavilov writes in his *Origin and Geography of Cultivated Plants*, 'Maize (*Zea mays*) plays the same role [in Southern Mexico] as wheat in the Old World: without it there would not have been any Mayan civilization'. It is included among plants of south Mexican and Central American origins, alongside kidney beans and runner beans. Corn was first domesticated in Mexico as many as 10,000 years ago, and cobs have been found in cave excavations in Tehuacán. Older varieties are smaller and multicoloured, as opposed to the more familiar yellow, and are more often dried and used as ornaments.

TOP *Plumeria* sp. (frangipani). Charles Plumier, who commemorated Fuchs in the naming of *Fuchsia*, was himself honoured with the genus *Plumeria*, a choice plant for Mediterranean and tropical gardens.

ABOVE Fuchs recorded the everyday use of *Ocimum basilicum* (sweet basil) in Germany, noting its regular appearance on windowsills.

William Turner

DATE	C. 1508–1568
ORIGIN	ENGLAND
MAJOR ACHIEVEMENT	*A NEW HERBALL*

William Turner is famous for producing the first scientific herbal
in English rather than Latin and thereby introducing medical and plant
knowledge to English speakers. When not studying or writing on
ornithology or botany, his life was spent in religious controversy. During
the English Reformation, his religious beliefs led to an exile in Europe,
where he travelled extensively and continued his studies. Turner described
himself as 'a physician delighting in the study of sacred literature';
today he is better known as the 'Father of English Botany'.

Acanthus mollis,
bear's breeches

Turner was born in Morpeth, Northumberland, in around 1508, and went up to Pembroke Hall, Cambridge University, in 1526, where he was awarded an MA in 1533. The second part of his famous work *A New Herball* is dedicated to his sponsor, Sir Thomas Wentworth: 'who hath deserved better to have my book of herbs to be given to him than he whose father with his yearly exhibition did help me, being student in Cambridge of physic and philosophy?'

Despite his achievements, Turner's life was one of turmoil, due to his passionate non-conformist beliefs. During the reign of Henry VIII, his writings were banned, and he was imprisoned for preaching without a licence. He was also twice exiled to Europe, and banned by the Church for non-conformity after returning home to Britain.

EARLY WORKS

During his first exile, Turner travelled throughout the Continent visiting France, then Italy, where he obtained a degree in medicine and studied under Luca Ghini, Professor of Botany at Bologna (and credited with creating the first herbarium). Turner then travelled to Switzerland and to Germany, where he met Leonhart Fuchs (see pp14–19). In Holland, Turner wrote that he became personal physician to 'the Erle of Emden for four full years' and in East Friesland, 'bought two whole Porpesses and dissected them'.

Despite his interest in plants, Turner's first printed book was on ornithology, *The Principal Birds of Aristotle and Pliny*, published in 1544 and including his own observations alongside those of his chosen authors. Turner notes, for example, that the blue tit is 'very fond of suet' and that in towns and cities, red kites are known for 'snatching food from children's hands'. Such is the importance of this work that Turner is often described as the 'Father of English Ornithology'.

On the subject of plants, Turner wrote *Libellus de re Herbaria Nova* in 1538, listing 144 plants with synonyms in Latin, Greek and English. He gave a forthright explanation of his reasons for writing the book in its preface: 'you will wonder, perhaps to the verge of astonishment, what has driven me, still a beardless youth, and but slightly infected with knowledge of medicine, to publish a book on herbary ... [I would rather] try something difficult of this sort rather than let young students who hardly know the names of their plants correctly to go on in their blindness.' Among the plants he mentions in the text are *Acorus calamus* (sweet flag) and *Acanthus mollis* (bear's breeches, pictured on p20).

This was followed ten years later by an extended version of the same work, entitled *The Names of Herbes*, which includes three times as many plants, and provides foreign terms alongside the English: 'Acanthus is called in greke acantha, in English Branke vrsin, in duche welshe bearenklawe, in frenche branke vrsine.... In the greatest plentie I euer sawe it, I did see it in my Lorde Protectour's graces gardine in Syon'. Syon Park is believed to have been the first botanical garden created in England, and the laying out of the formal gardens within the wall

around Syon House is traditionally attributed to Turner. He is also said to have planted the mulberry trees by the east front, which remain today. Remarkably, only one original copy of the *Libellus* and fewer than ten of *The Names of Herbes* survive.

Colchicum autumnale (meadow saffron) was observed and noted by Turner near Bath.

Turner's great work, *A New Herball*, arranged alphabetically and published in three parts between 1551 and 1568, broke with tradition, becoming the first important herbal to be written in English and the first scientific record of some 238 native species. Most of the illustrations came from his friend Leonhart Fuchs but they were poor copies. Turner attempted to analyse plants scientifically, describing their appearance and giving details of where they grew, together with medicinal and other uses. The text was written in English to enable ordinary people who could read, and apothecaries whose Latin was poor, to identify and use plants safely.

Turner's publications include references to plants seen by him on various research trips. There are descriptions of plants from his childhood: 'I never saw any plaine tree in Englande saving one in Northumberlande beside Morpeth and another at Barnwel Abbey beside Cambrydge'. In Somerset he saw meadow saffron 'growe in the west cuntre besyde Bathe'. He also provides an early description of the famous Glastonbury thorn: 'about six myles from Welles, in ye parke of G[l]assenberry there is an hawthorne which is grene all the wynter, as all they that dwell there about do stedfastly holde.' Further afield, he sighted several plants on 'Mount Appenine besyde Bonony', among them cyclamens and sumac, which the Italians used to tan leather, and in Chiavenna he found 'Monkshood growing in great plenty upon the alps'.

NEW NAMES FOR OLD

Where there were no common names, Turner created his own, often translated directly from the Latin. Some have religious overtones. Wood sorrel is called 'Alleluya, because it appereth about Easter when Alleluya is song agayn'; Solomon's seal 'is called in English scala celi' (stairway to heaven) because of its leaf formation. He experimented, too: bog myrtle 'is tried by experience that it is good to be put in beare [beer], both [by] me and by diverse other in Summersetshyre.' One event, however, illustrates the dangers associated with experimentation: 'I washed an aching tooth with a little opio mixed with water; and a little of the same unawares went down, within an hour after my handes began to swell about the wrestes, and to itch, and my breth was so stopped, that if I hand not taken in a pece of th roote of masterwurt ... with wyne I thynck that it would have kylled me.'

Turner created several gardens in his lifetime in addition to the one at Syon – including one near Tower Hill in London. A memorial to him can be seen close by, on the southeast wall of St Olav's church.

A hand-coloured woodcut featuring an illustration of a rose, taken from a 1568 publication of Turner's great work, *A New Herball*.

William Turner:
INSPIRATION FOR GARDENERS

✤ William Turner mentions that the hawthorn known as *Crataegus monogyna* 'Biflora' is unusual in that it flowers not only in May but in mid-winter, too. Tradition says that it flowers on Christmas Day, but depending on the weather, it blooms in milder periods any time from November until April. Legend tells that after the crucifixion of Christ, Joseph of Arimathea preached Christianity in England. Tired of travelling and being ignored, he rested on 'Weary-all Hill', at Glastonbury, where he prayed for a miracle to convince the doubters. The staff he leaned on, 'being thrust into the ground', simultaneously burst into leaf and flower. It was Christmas Day, and the miracle is said to be repeated every Christmas. Sprays are sent to the royal family as decoration for Christmas Day; the late Queen Mother placed hers on a writing desk and Queen Elizabeth II was said to have put hers on her breakfast table. This small, spreading, spiny deciduous tree or shrub likes moisture-retentive, free-draining soil, in sunshine or partial shade.

✤ When William Turner was Dean of Wells, in Somerset, he created a garden at what is now known as the Old Deanery, off Cathedral Green. The garden was restored to a Tudor layout and replanted by volunteers in 2003 using plants that were mentioned in *A New Herball*.

✤ One of the plants Turner mentions is 'perfoliata', a 'herbe wyth a leafe lyke pease & lytle blacke seedes in the top ... it may be called in England Thorowwax because the stalke waxeth thorowe the leaues'. *Bupleurum rotundifolium*, a cornfield annual, retains the same common name of thorow-wax. It looks like a spurge but is actually a member of the carrot family. Seed companies sell several selected forms, which are attractive in a contemporary cottage garden and excellent for flower arranging.

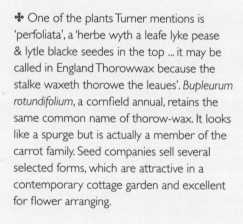

LEFT *Bupleurum rotundifolium* (thorow-wax) was given its common name by Turner, who first noted it growing in Germany.

RIGHT *Crataegus monogyna* 'Biflora' (Glastonbury thorn) was another plant observed by Turner in the wild and later described in his work.

John Ray

DATE 1627–1705	
ORIGIN ENGLAND	
MAJOR ACHIEVEMENT *HISTORIA PLANTARUM*	

A naturalist, theologian and prolific writer, John Ray published around
20 works, including a 1669 paper, 'Experiments concerning the Motion
of Sap in Trees', containing an early description of dendrochronology
– explaining how an ash tree could be aged by its rings. Ray's works were
based on his journeys, observing plants in their native habitats; Carl Linnaeus
would later draw on them for inspiration in creating his famous binomial
system. Known as the 'Father of Natural History', Ray published systematic
works on plants, birds, mammals, fish and insects, bringing order to the
chaotic mass of names and systems used by naturalists at the time.

Anethum graveolens,
dill

John Ray, son of an Essex blacksmith, proved that social class was no barrier to intellect by attending the local grammar school, and gaining a scholarship to Cambridge University aged 16. While there he studied at three colleges; lectured in Greek, Latin, mathematics and humanities; preached in the college chapel and became friends with John Nidd, Francis Willughby and Peter Courthope before leaving Cambridge in 1662, casting himself, he said, 'upon Providence and good friends' to fulfil his wish to travel, observing nature – particularly plants. His inspiration came in part from his mother, who during his childhood had shared her good knowledge on the uses of plants (such as dill, pictured on p26).

While at university, Ray had packed a small garden with plants he had collected or received as gifts from friends, analysing the differences between each one and developing his own system of classification. In doing so, he had discovered the limitations of existing botanical literature and became certain it could be improved. In the early 1650s he began to catalogue the plants he found during long horse rides through the Cambridgeshire countryside. His *Catalogus plantarum circa Cantabrigiam nascentium* (1660) described 558 wild species and crops growing in the county. Arranged alphabetically, it included the uses of the plants, alongside their English names and locations. Ray also identified several new plants and grouped similar plants together, rather than dividing them up (taxonomists are still divided into 'splitters' and 'lumpers'). His catalogue contained a brief outline of the structure and classification of plants, according to the principles of Swiss botanist Jean Bauhin, and revealed Ray's own meticulous approach to science.

A LIFE OF TRAVEL AND OBSERVATION

On 9 August 1658, Ray embarked on the first of seven extensive botanising trips throughout Great Britain, riding to Northampton, Warwick, the Peak District, Worcester and Gloucester, observing and documenting plants, nature and landscapes along the way. Inspired by what he saw, Ray decided to compile a work cataloguing every English plant, writing to friends across the country to ask for their help. As a result, he made seven extensive tours of Britain, accompanied by his university friends, making meticulous observations of flora, fauna and sea creatures (on one occasion dissecting a porpoise). He also made a mammoth three-year trip around Europe, at Willughby's instigation, which allowed Ray to collect and catalogue non-native plants along the way.

After his extensive travels through Britain, Wray published his *Catalogus Plantarum Angliae* in 1670, dedicated to Willughby. A second edition was published in 1677, with an appendix of newly observed plants by 'John Ray and his friends'. The preface proclaimed Ray's conviction that the divine creation of plants indicated there was a reason for their existence, which could only be discovered by putting them to practical use. Continuing his work on the classification and nature of plants in *Methodus Plantarum Nova* (1682), 'a catalogue of Species not native to England', Ray classified plants using the whole plant: 'There was leisure

to contemplate by the way what lay constantly before the eyes and were so often trodden thoughtlessly under foot ... the shape, colour and structure of particular plants fascinated and absorbed me: Interest in botany became a passion.' Ray also believed that local species were likely to be preserved by providence somewhere else in the world – that the Creator would ensure the survival of a species by placing it in different locations around the world – and was first to make a distinction between monocotyledons and dicotyledons.

Ray's greatest work was his *Historia Plantarum*. The two volumes, published in 1686 and 1688, described approximately 6,100 species of plants. The second volume also contained 15 rare North American plants from bishop and botanist Henry Compton's garden (see pp38–43). A third volume, detailing a further 10,000 species, was based on research rather than Ray's own observations, old age and infirmity preventing him from visiting Oxford University and the Chelsea Physic Garden, where he could have observed new and exotic plants first hand. The classification used in the *Historia Plantarum* was an important step towards modern taxonomy, and volume one included the first usage of the term 'species'.

Salvia officinalis (common sage) was widely grown in the 1600s for its culinary and medicinal uses.

In 1690 Ray published *Synopsis Methodica Stirpium Britannicarum*, following the classificatory system of his *Methodus Plantarum*. With botanist Sir Hans Sloane's encouragement, he drew up all but one of the county lists of plants for Edmund Gibson's edition of Camden's *Britannia* (1695), and he issued a further catalogue of European plants (*Stirpium Europaearum Extra Britannias Nascentium Sylloge*) in 1694. This included a stout defence of his method of classification against that of German botanist Augustus Rivinus (also known as A. Q. Bachmann), based on differences in the shape of flowers. An exchange of letters appeared in the second edition of his *Synopsis*, where Ray referred to the writings of French taxonomist Joseph Pitton de Tournefort, and he took the work of both Rivinus and Tournefort into account in *Methodus Plantarum Emendata et Aucta* (1703), when he revised his classifications, introducing divisions between flowering and non-flowering plants. This time of turmoil for plant classification was only resolved once Linnaeus's binomial system became established.

In 1691 Ray published *The Wisdom of God Manifested in the Works of the Creation*. Based on his sermons at Trinity College, this was a statement of his faith. He believed in 'natural theology', in which the heavens, geology, botany, zoology and human anatomy were evidence of a providential, benevolent god. Ray believed 'Divinity is my Profession', stating, 'I know of no occupation which is more worthy or more delightful for a free man than to contemplate the beauteous works of Nature and to honour the infinite wisdom and goodness of God the Creator'.

On 15 March 1679, upon the death of his mother, Ray and his wife moved back to Black Notley, his birthplace, and settled in the family home. Suffering from chronic ill health during his later years, he died at home on 17 January 1705, having laid the foundations for modern taxonomy.

John Ray:
INSPIRATION FOR GARDENERS

✤ John Ray was the first to formally recognise the distinction between monocotyledons and dicotyledons, noting differences in their leaves, stems, roots and flowers. Monocots have one seed leaf; dicots have two. Monocots tend to have fibrous roots, which are found near the soil surface, while dicots have one main root, the taproot, with smaller roots branching from it. As monocots grow, the sap system carrying nutrients to each portion of the plant is scattered; in dicots the system is structured. Their leaves look different, too; monocot leaves have parallel veins, but dicots form branching veins. The flowers of monocots are grouped in threes, but dicotyledon flowers are in groups of four or five. Monocot pollen has one furrow; in dicots there are three.

✤ Ray learned about culinary and medicinal uses of plants from his mother, Elizabeth, and some of these are still familiar today. Examples include the use of *Salvia officinalis* (sage) and *Thymus serpyllum* var. *albus* (white-flowered creeping thyme) to flavour stuffings, and *Anethum graveolens* (dill) as an accompaniment to fish.

BELOW *Thymus serpyllum* var. *albus* (white-flowered creeping thyme). Ray was introduced to thyme by his mother, who passed on her knowledge of plants.

BELOW A medicinal plant in Ray's era, *Campanula persicifolia* (fairy bellflower) is now used as an ornamental herbaceous plant.

❧ *Campanula persicifolia* (fairy bellflower) was known to Ray as a remedy for mouth ulcers and sore throats. Now an ornamental, it thrives in fertile, moist, well-drained, neutral to alkaline soil, in sun or part shade. Used as a dressing for wounds in Ray's era, the late-spring flowering *Polygonatum multiflorum* (common Solomon's seal) is now used as a garden plant, in moist, free-draining, organic rich soil, in part shade.

❧ *Rosa* 'John Ray' was bred by Heather Horner and named in 2000 at the request of the John Ray Trust, in Braintree, Essex, to mark the millennium. It is a deep yellow, semi-double, cluster-flowering rose, with a mild to strong fragrance, producing flushes of blooms throughout the season. This rarely found rose grows at Middleton Hall near Tamworth in Staffordshire, home to Francis Willughby and where John Ray lived between 1660 and 1673. As with all roses, 'John Ray' is a surface-rooting plant, so avoid hoeing and control weeds by mulching around it in spring with a thick layer of well-rotted organic matter.

TOP Ray mentions *Cerinthe major* 'Purpurascens' (honeywort) in his *Historia Plantarum*. It is still popular as a garden plant, for cut flowers.

ABOVE The spring-flowering *Polygonatum multiflorum* (common Solomon's seal) was known to Ray as a good dressing for wounds.

Mary Somerset

DATE 1630–1715	
ORIGIN ENGLAND	
MAJOR ACHIEVEMENT 12-VOLUME HERBARIUM OF PRESSED PLANTS	

Mary Somerset, Duchess of Beaufort, was a passionate botanical collector, cataloguer and cultivator of exotic plants in her gardens and greenhouses in London and Badminton, which were among the finest of their day. She kept detailed lists and records of her plants and her collections, and expanded her botanical knowledge by reading and corresponding with the great botanists and gardeners of the day, including John Ray and Sir Hans Sloane. Her greatest legacy is a 12-volume herbarium of pressed plants that she bequeathed to Sloane after her death.

Bombax ceiba,
South American
silk cotton tree

M ary Somerset, first Duchess of Beaufort, was well placed to garden and botanise. Not only did she have an estate in Gloucestershire but her neighbours in Chelsea were physician and collector Sir Hans Sloane and the Chelsea Physic Garden. The Duchess's skill as a gardener, combined with her wealth and personal connections, enabled her to accumulate a diverse collection of botanical curiosities. Plant collector James Petiver wrote: 'in her grace the Duchess of Beaufort's most Noble Garden as the Matchless Stoves at Badminton in Gloucestershire, I the last summer met with many new rare and very curious plants, most of them raised to that perfection I never saw before.' Like many of the great estates of the time, it included 'an extensive orchard of many different fruits [including oranges under glass] as well as gardens containing fashionable flowers of the time: tulips, narcissus, ranunculus, anemones, etc.' There were also collections of 'curios' at her London home, Beaufort House. During the last decade of the seventeenth century, the Duchess amassed a collection of variegated plants in pots, including 'Strip't Phillerea, (*Phillyrea latifolia* 'Variegata'), Strip't thyme (*Thymus serpyllum* 'Aureus') ... Silver Strip't Myrtle (*Myrtus communis* 'Variegata') and Silver Strip't honeysuckle (*Lonicera periclymenum* 'Variegata')', which were placed in prominent places in the knot garden or wilderness. Her lists from 1692 to 1699 feature an increasing range of 'striped plants'. The first contains 28 varieties, mainly evergreens and herbs; by 1697 these numbered 69, including deciduous woody plants, such as beech (*Fagus sylvatica* 'Luteovariegata') and netted elder (*Sambucus nigra* 'Marginata'), and a greater variety of herbs and ornamentals like sweet williams and sage. A third list contains around 91 varieties, including several *Amaranthus*: '*A.tricolor* ye common', the cultivar 'Splendens' ('from ye East Indiis, much better'), and another that had 'a deep red spot in ye middle of ye leav's a very good kind'.

A GLOBAL COLLECTION

The Duchess's hot houses were renowned for botanical curiosities that had come from all parts of the world, notably India, Sri Lanka, Japan, China, South Africa and the West Indies. George London, of the Brompton Park Nursery, and William III's gardener at Hampton Court, was a major source of seeds and plants. In 1695 and 1696 the Duchess received a number of letters from him containing seeds from the West Indies, notably the Barbardos lily (*Hippeastrum puniceum*); he also sent seeds from Guinea, the Cape of Good Hope, Portugal, the Canary Islands and from the East Indies – the 'Java bean tree', black datura and tamarind. A letter accompanying a 'Cheste of silke cotton trees' (*Bombax ceiba*, pictured on p32) included a list of the merchant ships (and their captains) on which these plants were travelling. These are later recorded in the Duchess's census of 1 June 1697 as growing in her garden, along with a 'Benjamin tree', or spice bush (*Lindera benzoin*), bamboo and a date tree.

The Duchess compiled detailed inventories of her plants, particularly during the last decade of the seventeenth and the first of the eighteenth century. These were

cross-referenced to the botanical authorities of the time, arranged alphabetically according to the country of origin and often accompanied by practical details, such as cultivation and propagation. ('Mark't in my booke', she writes on a list of plants from Portugal and Guinea in 1695.) From around 1700, the botanist William Sherard was employed at Badminton as tutor to Mary's grandson. While there, through his correspondence and contacts, the Duchess added over 1,500 plants to her collection.

STUDENT AND PATRON

The Duchess exchanged plants and knowledge with all of the great gardeners and botanists of her day; Sloane also gave her the transactions of the Royal Society to read, indicating the depth of her understanding of botany and horticulture. It is evident from the cross-references on her inventories that she also read the works of luminaries of the time, including John Gerard, John Parkinson and John Ray (see pp26–31). However, she was not averse to criticising the work of established botanists, based on what she considered more accurate information, gleaned from the plants she was growing herself, stating, 'none of the Lyrium in Parkisons H[erbal] are like those plants of mine', and declaring others 'not well described by any of them the plants I have being more beautifull than any of these prints'. She corrected another work, writing, 'Wild Callabash and Wild Cashe trees, should have been with the trees', and in another note – 'These plants I grow which I can find no figures of in any of my books, they are therefore to be described as they grow now at Badminton in 1693' – indicated that some of her plants were yet to be recorded and may have been new species to cultivation.

Several cultivars of *Amaranthus tricolor* (tampala) were grown and recorded by Mary Somerset. She enjoyed growing unusual tropical plants.

Although her vast plant collections and great stoves have long disappeared, her legacy, a 12-volume large-folio herbarium, containing dried specimens of 'plants, most rare and some common gathered in the field and gardens at Badminton, Chelsea, etc. dryed by order of Mary Duchess of Beaufort', and entitled 'the Duchess of Beaufort's plants', was left to Sir Hans Sloane on her death and is now in the Natural History Museum. The Duchess also oversaw a two-volume *Florilegium* of botanical illustrations of some of the more interesting plants in her collection. The first volume was painted by Everhardus Kickius, and the second volume by Daniel Francome, both of which are held in the library at Badminton House. We know that Sloane loaned the Duchess's herbarium to botanist John Ray, who made various annotations on its pages and a few revisions to her descriptions, indicating that he respected the Duchess's work and its careful preservation. Sloane also obviously shared these views. The Duchess of Beaufort died on 7 January 1715. She was suitably commemorated a century later with the naming of *Beaufortia*.

Mary Somerset:
INSPIRATION FOR GARDENERS

✤ Among the plants painted in the Duchess's *Florilegium* is *Melianthus major* (honey flower), an evergreen sub-shrub introduced from the Cape of Good Hope to Britain in 1688. Its grey-green leaves with serrated edges are very architectural but they have an unpleasant smell. Flowers, produced during long, hot summers, are dark red in colour, with copious black nectar that can be shaken from the flower onto the palm of the hand. It is then that you discover why it was given its common name. Honey flower needs moderately fertile soil in a sheltered, sunny position, away from cold or drying winds, and a dry mulch in winter to protect the roots. In colder areas, grow it in a pot and overwinter under protection.

BELOW *Melianthus major* (honey flower), illustrated in the Duchess's *Florilegium*. It has spectacular spikes of flowers in hot summers.

✤ In volume two of the *Florilegium*, Francome painted *Hibiscus rosa-sinensis* (Chinese hibiscus), now a widespread garden shrub throughout the tropics, grown for its large, brightly coloured flowers. It can be grown under glass or as a houseplant in cooler climates in a warm bright position, with moderate to high humidity; mist around the plant with tepid water or stand it on a tray of pebbles filled with water to the bottom of the pot. You should water this plant when the compost surface starts to dry out, but reduce watering in winter. The double-flowered 'Chiffon' range of hardy hibiscus come in lavender, pink and blue. These hardy plants are covered with flowers from mid-summer to autumn.

✤ The Duchess grew pelargoniums that had arrived from South Africa. Among them were those with scented leaves. *Pelargonium papilonaceum* has a lemon fragrance, 'Prince of Orange' smells of orange, 'Lady Plymouth' of rose, and the large furry leaves of *Pelargonium tomentosum* smell of peppermint. Pelargoniums are not frost hardy, so are better grown in containers and brought indoors over winter before the first frosts. Reduce watering from early autumn. Feed during the growing season with high-potash fertiliser for compact, flower-covered plants, and grow in fibrous, free-draining compost.

TOP RIGHT *Hibiscus rosa-sinensis* (Chinese hibiscus) appeared in volume two of Somerset's *Florilegium*. It is a garden plant in warmer climates and a house plant in cooler conditions.

RIGHT The Duchess loved pelargoniums and maintained a large collection. Among this was *Pelargonium tomentosum* (peppermint geranium).

Henry Compton

DATE	1632–1713
ORIGIN	ENGLAND
MAJOR ACHIEVEMENT	BOTANIST AND BISHOP

After a short career as an officer in the Life Guards, Compton turned to religion and, as Bishop of London, was responsible for the Church of England overseas, including the colonies in America, the Caribbean, Africa and India. When he appointed clergymen, those with a botanical interest were always favoured. After being suspended by King James II for his overly Protestant beliefs, Compton focused on bringing glory to his garden by planting rare and unusual plants. In doing so, the 'Botanising Bishop' became one of the great ecclesiastical botanists of all time.

Magnolia virginiana,
swamp bay

Τ he garden at Fulham Palace was already well established in 1675, when Henry Compton arrived to plant and garden the 36 acres of land, but it had yet to reach the peak of its fame. A biography published after his death recorded that he 'had a great genius for Botanism, and ... he apply'd himself to the improvement of his Garden at Fulham, with new variety of Domestik and Exotick plants'; contemporary botanist Richard Pulteney also noted Compton combined 'his taste for gardening' with 'a real and scientific knowledge of plants'. According to an 1813 book on the progress of English botany, Compton's aim was 'to collect a greater variety of Greenhouse rarities and to plant a greater variety of hardy Exotic trees and Shrubs than had been seen in any garden before in England', and in many ways he succeeded.

Compton appointed George London (subsequent founder of the Brompton Park Nursery) as his advisor and gardener, and by 1686, William Penn's gardener was hoping for an arrangement with the 'Bishop of London's gardener' to supply him with garden plants in exchange for products from Pennsylvania.

SOURCING THE NEW WORLD

Compton sent Reverend John Banister, one of the first university-trained naturalists in North America and the first notable Virginian botanist, to North America via the West Indies in 1678, where, botanist Richard Pulteney noted, Banister 'industriously sought for plants, described them and drew the figures of the rarer species'. Banister sent trees, shrubs and seeds back to Fulham Palace and other gardens, including the University of Oxford Botanic Garden and the Temple Coffee House Botany Club (both Compton and London were members), where the most eminent botanists of the day met to compare notes once a week. In the second volume of his 1688 *Historia Plantarum*, John Ray (see pp26–31) attempted to describe all known plants, including those from Virginia growing in the Bishop's garden, making it the first publication dealing with North American flora. Compton was a friend and supporter of Ray.

Banister's first consignment (1683) included *Liquidamber styraciflua* (sweet gum) and *Lindera benzoin* (spice bush). The second (1687) included *Acer negundo* (box elder) and *Magnolia virginiana* (the first magnolia to be grown in Britain – pictured on p38). He also sent *Gillenia trifoliata* (Bowman's root), which was successfully grown by Head Gardener William Milward. It was named after Arnold Gille (Gillenius), a seventeenth-century German botanist, and by 1713 was already found in several gardens in London. Banister's life was cut short when he was accidentally shot dead (or fell from a rock – it remains a mystery) while on expedition. Compton also received plants from others. Reverend John Clayton, a rector in Jamestown, and Sir William Berkeley, Govenor of Virginia, sent regular shipments of trees to England, including at least one recorded shipment to Fulham Palace. Mark Catesby, an English naturalist who had moved to Virginia and become a professional plant collector, also sent seeds and rare plants to Compton shortly before the Bishop's death.

Plants arrived from other parts of the world, too. Scottish surgeon and botanist James Cunningham brought a large *Rhus chinensis* (Chinese gall), which contemporary botanist E. H. M. Cox said had 'ripe berries on it from China, which James Petiver [plant collector and "compulsive natural historian"] gave to the Right Reverend, my Lord Bishop of London'. Compton's garden was so esteemed that when new plants arrived, some were always sent to Fulham Palace. In 1687 the collector Hans Sloane sailed to Jamaica as physician for the Duke of Albermarle, describing every plant he saw in detail. On his return, he brought with him around 800 plants, which were sent to Compton and Mary Somerset, the Duchess of Beaufort (see pp32–37).

Bishop Compton dispersed his clergy worldwide to preach and collect plants, such as *Crotalaria juncea* (bengal hemp).

PLANT COLLECTOR EXTRAORDINAIRE

Hortus Kewensis credits Compton with over 40 new introductions, two-thirds of them hardy trees and shrubs. He had a great passion for trees; in the 38 years that he presided as Bishop, the grounds became densely planted with *Liquidambar*, *Liriodendron*, scarlet oak, black walnut, pines, juniper, cedars, maples, hollies, hickories, hawthorns, dogwoods, robinias and more. Professor John Hope, Professor of Botany at Edinburgh University, visited in 1766, and was the first to note a 20m (65½ft)-high *Juglans nigra* (black walnut), a *Quercus suber* (cork oak) of 12m (39ft), and an *Acer rubrum* (red maple) of around 9m (29½ft). The garden also included *Lycium barbarum* (bearing goji berry fruits), *Adiantum pedatum* (five-fingered maidenhair fern), *Arisaema triphyllum* (bloody arum), *Dodecatheon meadia* (shooting star), *Mertensia virginica* (Virginian bluebells) and several pelargoniums, including *P. capitatum*, *P. myrrhifolium* and *P. cucullatum*, all introduced in 1690 by Hans William Bentinck, Head Steward of William of Orange (who Compton supported). *P. inquinans*, one of the parents of the modern bedding geranium, also featured in the garden; the other parent, *P. zonale*, was grown by the Duchess of Beaufort.

Arisaema triphyllum (bloody arum), a native of the eastern North American woodlands, was recorded as growing in Compton's garden.

There was space for edibles, too. Compton was interested in vegetables and grew kidney beans to eat at a time when they were seen as purely ornamental. Many of the names of his plants have now changed – the *Clematis hederacea* listed in the *Sloane Herbarium*, for example, is actually *Campsis radicans*. It is not known whether Compton grew the 'Fearne tree of Virginia' (later named *Comptonia asplenifolia* in his honour), but it was already in cultivation in his day, being grown by the Duchess of Beaufort and in the Chelsea Physic Garden before 1700.

According to William Watson's *Philosophical Transactions*, most of Compton's plants were dispersed to other collections after his death by 'ignorant persons ... to make way for the more ordinary productions of the kitchen garden'. Some plants went to the University of Oxford Botanic Garden, others to the Chelsea Physic Garden, and the Head Gardener sold yet others to nurserymen, which were then dispersed into general cultivation. Fulham Palace garden underwent a major restoration in recent years.

Henry Compton:
INSPIRATION FOR GARDENERS

✤ Bishop Compton grew *Rhododendron viscosum* (swamp azalea), the first azalea to be grown in Britain. It makes an excellent garden shrub due to it being late flowering, and producing exquisitely fragrant white to pink-flushed flowers from late spring into summer. It also has excellent autumn colour. It is said to be a parent of a great number of azalea hybrids; Loddiges' nursery catalogue of 1836 lists 107 varieties. Grow this plant in pots of ericaceous compost, and water it with rainwater, or plant it in acidic soil in a sheltered position, in dappled shade or sunshine (but not scorching sunshine). As a native of swampy lowlands, it is best grown in acidic, woodland soil. It tolerates moist to wet soils, but will not grow in soils where the roots are permanently underwater. Deadhead and feed after flowering, and mulch with well-rotted compost in spring.

✤ *Liquidambar styraciflua* (sweet gum) is a native of the eastern United States, and also often found in swampy ground. It has long been valued for its stately form and handsome foliage. There are several good selections, including 'Lane Roberts', which produces outstanding black-red, ember-like autumn colour. It was named after eminent London gynaecologist, rugby player and keen gardener Cedric Sydney Lane-Roberts, who purchased several seedlings from Hillier Nurseries. When Sir Harold Hillier visited one day, he was impressed with one of the seedlings, asked for some scionwood and named the new cultivar after his friend.

BELOW Compton grew *Rhododendron viscosum* (swamp azalea). This was the first azalea species to be grown anywhere in Britain.

✤ *Lycium barbarum* (Chinese box thorn), grown by Compton, produces goji berries – now considered a 'super food'. It is a fast-growing deciduous shrub with long, lax stems. It has long been grown in gardens and has naturalised in many parts of the world, including Britain. Its white and purple, trumpet-shaped flowers appear from early spring to midsummer, and are followed by the shiny red berries. These can either be picked and eaten straight from the bush, or dried. It takes two to three years for the shrub to produce its first crop.

TOP RIGHT *Lycium barbarum* (Chinese box thorn), which produces goji berries, grew in Compton's garden. Native to south-eastern Europe and Asia, the plant has now been naturalised throughout Britain.

RIGHT Compton filled his grounds with trees, including *Liquidambar styraciflua* (sweet gum). This can be mistaken for a maple, due to the shape of its leaves and its vivid colours in autumn.

William Dampier

DATE 1651–1715	
ORIGIN ENGLAND	
MAJOR ACHIEVEMENT *A NEW VOYAGE ROUND THE WORLD*	

William Dampier, a maritime maverick, combined the unlikely disciplines of exploration, piracy and natural history. He circumnavigated the world three times, collected plants from Australia 71 years before Banks arrived and published detailed records bringing these experiences to life. Dampier was the first to describe the breadfruit in English, described the Galápagos turtles before Darwin and even produced the first recipe in English for guacamole. His reports on breadfruit also led to the voyage of HMS *Bounty*, which infamously ended in mutiny in 1789.

Abrus precatorius,
coral bead plant

O n leaving school, Dampier became apprenticed to a seaman in Weymouth, fought briefly during the Third Anglo-Dutch War, was employed on a Jamaican sugar plantation, then traded in dye in Mexico. His pay was so poor, that Dampier became a 'privateer', raiding Spanish towns along the coast of southeast Mexico. By 1678, he had made enough money to return to England and marry. However, he returned to Jamaica, and piracy, a year later, becoming infamous for his attacks on the Spanish along the east and west coasts of Central America. From March to May 1686, Dampier navigated from Mexico across the Pacific to Guam, exploring the South China Sea, the coasts of Southeast Asia, Western Australia, Sumatra and India, arriving back in England in September 1691, the first seventeenth-century Englishman to circumnavigate the globe. Dampier capitalised on his notoriety and success by publishing *A New Voyage Round the World* (1697), giving a detailed description of the flora, fauna and people he encountered on his travels. In 1699, the Admiralty gave him command of HMS *Roebuck* to circumnavigate Terra Australis (Australia) to assess its potential; once again he published an account of his voyage – *A Voyage to New Holland* (1703). From 1703 to 1707 he returned to harry the Spanish in the Pacific, this time unsuccessfully. Dampier left on his third circumnavigation in 1708, with two ships, the *Duke* and the *Duchess*, returning to England in 1711. This time in triumph and wealthy enough to retire.

DAMPIER THE NATURAL HISTORIAN

On his second circumnavigation, on 22 February 1699, Dampier left the Cape Verde islands in HMS *Roebuck* for Brazil, arriving off the coast of Bahia, which he described as 'checker'd with Woods and Savannahs'. After anchoring at Salvador, the historic capital of colonial Brazil, he began botanising around the city, noting that it was 'well watered with Rivers, Brooks and Springs.... The Soil in general is good, naturally producing very latge [sic] Trees of divers sorts, and fit for any uses. The Savannahs also are loaden with Grass, Herbs, and many sorts of smaller Vegetables', but noting, too, that gardens were 'order'd with no great Care nor Art'.

The specimens Dampier collected were usually just fragments of larger plants; the largest of his Brazilian specimens could be laid on a single sheet of paper. At least two specimens from his Brazilian collections appear to be missing from the herbarium at Oxford University, where 66 that are attributed to Dampier remain (27 from the Americas, 27 from Australia, 9 from Southeast Asia and 3 unknown). Two sterile shoots, 'from the new Island in the East Indies discovered by Dampier', (Australia) are in the Sloane Herbarium at London's Natural History Museum. Dampier also recorded the names and detailed information on the uses of 73 plants encountered in Brazil. He used English names where possible, but if unavailable, he informed his readers, they were recorded 'as they were pronounc'd to me'. In five cases, the details of the plants and their names were reported to him by 'an Irish Inhabitant of Bahia', rather than from his own observations.

AN ENGLISHMAN IN AUSTRALIA

In August and September 1699, William Dampier became the first Englishman to collect plants in Australia. He passed Dirk Hartog Island to enter Shark Bay, on the coast of what is now Western Australia. He then visited an island in the Dampier Archipelago that he called 'Rosemary Island' – because a plant there, *Olearia axillaris*, reminded him of rosemary back home (the shape of the leaves was similar, although it was not fragrant and turned out to belong to a different family). He then visited Lagrange Bay. Dampier collected and pressed 26 species, mostly from Dirk Hartog Island, including *Solanum orbiculatum*, endemics including *Beaufortia sprengeliodes* (originally *Beaufortia dampieri*, pictured on p49) and reflexed panic grass (*Paractaenum novaehollandiae*). All are now found in the Oxford University Herbaria, but were unknown to scientists before they were collected by Dampier. According to him, in Shark Bay 'the blossoms of the different sorts of trees were of several colours, as red, white, yellow etc. but mostly blue'. So it is appropriate that the plant commemorating him, *Dampiera incana*, has blue flowers; it was named by Robert Brown, first Keeper of Botany at the British Museum in London.

Dampier also pressed the flowers of a spectacular plant from East Lewis Island in the Dampier Archipelago, known as desert pea or the Dampier pea, neglecting the leaves. It was first named *Clianthus formosus*, then *Swainsona formosa*, but was recently proposed as *Willdampia formosa*.

Dampier reported that 'many of my Books and Papers [were] lost' on the journey home, when his ship sprang a leak and sank off Ascension Island. However, he managed to save his specimens (possibly the only herbarium specimens to survive a shipwreck), which were then exposed to the elements for several weeks before the crew were rescued. We also still have Dampier's recordings of many newly discovered plants in his journals, such as *Crotalaria cunninghamii*, *Canavalia rosea* and *Abrus precatorius* (pictured on p44).

Dampier collected the first specimens of *Swainsona formosa* (desert pea); these can be found today at an Oxford University herbarium.

On his return, Dampier passed all of his specimens to Dr John Woodward, Professor of Physic at Gresham College, London, and Fellow of the Royal Society of London, who lent specimens to John Ray (see pp26–31). Ray named 18 of them in the appendix to the third volume of his *Historiae Plantarum* (1704). He also lent some to Leonard Plukenet, botanist to Queen Mary II, who described and illustrated them in the appendix of his *Amaltheum Botanicum* (1705). Dampier also described some of his plants in his own publication *A Voyage to New Holland*, with line drawings of 18 plants, including 2 seaweeds, of which 9 were Australian plants.

After a life of plants, travel and adventure, Dampier retired to the parish of St Stephen's, Coleman Street, London, where he lived modestly on his pension and earnings from privateering and died in 1715.

William Dampier:
INSPIRATION FOR GARDENERS

✤ Among the species Dampier collected was *Scaevola crassifolia* from Dirk Hartog Island, a drought-tolerant coastal shrub with blue fan-shaped flowers. The plant, with tiny seeds, is available commercially and although not frost hardy, can be grown in a container and overwintered indoors, or grown as a border specimen or hedging in warmer climates. Replicate its natural habitat by growing it in sandy or free-draining soil, in full sunshine, and deadhead to prolong the flowering season. In Australia there is a low-growing selection called 'Flat Fred'. A close relative, the *Scaevola aemula* 'Blue Wonder' (fairy fan-flower) and similar cultivars are also good subjects for pots and hanging baskets in cooler climates.

✤ Dampier discovered *Swainsona formosa*, (desert pea, formerly Sturt's desert pea). It was named after English explorer Charles Sturt, who, in his *Narrative of an Expedition into Central Australia* (1849), refers to the flowers several times, and to the harsh habitat where he found the plant. Sturt saw it in large numbers while exploring central Australia in 1844 and wrote that, beyond the Darling River, 'we saw that beautiful flower the *Clianthus formosa* [sic] in splendid blossom on the plains. It was growing amid barrenness and decay, but its long runners

BELOW Dampier was the first to bring *Scaevola crassifolia* to Britain, from Dirk Hartog Island. Its relative, *Scaevola aemula* (pictured), is now a popular choice for containers.

ABOVE Among the species collected and pressed by Dampier was the shrub *Solanum orbiculatum*, a native of Western Australia.

were covered with flowers that gave a crimson tint to the ground'. The flowers are usually blood red or scarlet, sometimes with a glossy black swelling or 'boss' at the base of the upper petal, but they can be white to deep pink. A bicoloured form is occasionally found in scarlet and white. Seeds remain viable for years: abrade the seed gently with sandpaper, carefully 'nick' the seed coat opposite the 'eye' with a sharp knife, or soak in warm water overnight. Plants grow happily in free-draining soil and sunshine, tolerating slight frosts once established. Sow seeds in pots of gritty compost and transplant them before the long tap root develops, or grow them in tall pots.

ABOVE *Beaufortia sprengeliodes* was originally named *Beaufortia dampieri* to honour the botanising buccaneer. The plant thrives in poor, free-draining soils.

John Bartram

DATE	1699–1777
ORIGIN	UNITED STATES OF AMERICA
MAJOR ACHIEVEMENT	PLANT SUPPLIER

Bartram, a farmer turned botanical collector, travelled widely, enduring arduous conditions to supply his friend, agent and fellow Quaker, Peter Collinson, with new plants – the first to arrive in Europe from North America. Collections of seeds arrived regularly in containers as aristocratic gardeners competed with each other to be the first to grow the new species and planting the landscape with American trees became fashionable. Linnaeus described Bartram as the 'greatest natural botanist in the world', and he was appointed Royal Botanist for North America by George III.

Quercus x heterophylla,
Bartram's oak

John Bartram was born on 23 March 1699 on a farm at Marple, Pennsylvania, the eldest son of William Bartram, who emigrated as a child from Compton, Derbyshire, and the first of the family to be born in the New World. A letter he wrote in 1764, reveals: 'I had always since ten years old, a great inclination to plants, and knew all that I once observed by sight, though not their proper names, having no person nor books to instruct me.' He focused on medical botany and later in life, between his duties as a farmer, developed America's first botanical garden on 8 acres of his own land, using plants he collected while on extensive trips throughout the eastern North American colonies. His travels, by boat, on horseback and on foot, took him north to New England, south to Florida and north to Lake Ontario. Sometimes he would collect for a few days, at other times for weeks or months, either alone or with his son William, who also became a notable naturalist, ornithologist and author.

SUPPLYING THE ENGLISH COLLECTORS

Throughout his life, Bartram's Quaker connections were influential. Merchant Joseph Breintnell and physician Samuel Chew both recommended Bartram as 'a very proper person' to assist fellow Quaker Peter Collinson, a London-based cloth merchant and keen collector of plants and seeds, in his quest to supply flora, fauna and exotic specimens of natural history from North America. The two began corresponding in 1732, exchanging several letters a year until Collinson's death in 1768, becoming best friends, even though they never met. Bartram agreed to send Collinson plants for a fee of five guineas a box, but Collinson, who expected to receive only a few boxes, had not anticipated his diligence and skill. Box after box of rare and fascinating plants arrived in London and Collinson was delighted. In return, Bartram received advice, encouragement, money and a constant supply of natural history books, which he enjoyed. Most of Bartram's longer trips were made in the autumn, after harvest, when he would gather ripe seeds and nuts, roots and bulbs.

Word of his collecting soon spread and Collinson became Bartram's chief agent in London. The containers he sent, generally containing 100 or more varieties of seeds, and occasionally herbarium specimens and natural history curiosities, became known as 'Bartram's Boxes'. Collinson selected his patrons shrewdly, to secure regular funding that could support future expeditions. It came from three main sources: Lord Petre agreed ten guineas per annum and became the foremost collector of North American trees and shrubs in Europe; a further ten guineas were supplied equally by the Duke of Richmond at Goodwood House, and Philip Miller of the Chelsea Physic Garden, where many of Bartram's early plants were grown (he later became advisor on American plants to the aristocracy). Bartram ultimately filled orders for 21 nursery gardeners and 124 individual customers, and he exchanged plants with at least 33 friends and correspondents. Competition became fierce. The Duke of Richmond urged Collinson to act quickly on his behalf or

'the dukes of Norfolk and Bedford will sweep them all away'. Others contributed occasionally, including Charles Hamilton at Painshill Park in Surrey, which now holds the John Bartram Heritage Collection of North American trees and shrubs in Britain. All painted their landscaped grounds with North American trees. Bartram both fuelled and benefited from the enthusiasm for landscape gardening that was sweeping England at the time; the nobility wanted exotic plants and trees for these gardens, and Bartram was the collector who supplied them.

Through his letters, Collinson introduced Bartram to many important naturalists of the day, both in England and in the colonies, including Dr John Fothergill, a distinguished English physician, authority on herbs and a Quaker. Carl Linnaeus also wrote to Bartram, requesting botanical information; Peter Kalm, one of Linnaeus's favourite students, travelled to America specifically to talk with Bartram, who became a sought-after and respected contact.

In 1765, Bartram's son William discovered *Franklinia alatamaha* (Franklin tree). It was last found outside of conservation in 1803.

In spring 1765, after Collinson's persistent lobbying efforts through the Duke of Northumberland, Bartram was appointed botanist to George III, with an annual stipend of £50. Bartram's seeds and plants now went to the Royal Botanic Gardens at Edinburgh and Kew.

This new and abundant influx of plants was of keen interest to both scientists and artists. Philip Miller was also able to clarify uncertainties in his *Gardener's Dictionary* using specimens he received from Bartram, as did Linnaeus, while respected botanical illustrators Mark Catesby and G. D. Ehret produced eagerly awaited illustrations of the American plants that thrived in Collinson's and Miller's gardens.

COLLECTORS

Most of Bartram's plant discoveries were named by botanists in Europe. He is best known today for the discovery and introduction into cultivation of a wide range of North American flowering trees and shrubs, including kalmia, rhododendron and magnolia species; for introducing *Dionaea muscipula* (Venus fly trap) to cultivation; and for the discovery of *Franklinia alatamaha* (the Franklin tree) in southeastern Georgia in 1765; it was later named by his son William. Bartram is remembered in a genus of mosses, *Bartramia*, and in plants like *Amelanchier bartramiana* (shadbush), *Commersonia bartramia* and *Quercus × heterophylla* (Bartram's oak).

In 1769 Bartram was elected to the Royal Swedish Academy of Sciences, having assisted Swedish naturalist Peter Kalm on a visit to North America. In 1772 he received a gold medal from the Society of Gentlemen in Edinburgh, and artefacts he sent to the Royal Society can still be seen at the British Museum. Bartram remained alert and productive until his death on 22 September 1777, allegedly precipitated by his concern for the safety of his beloved personal garden in the face of advancing British troops.

John Bartram:
INSPIRATION FOR GARDENERS

✤ Bartram's oak (*Quercus* × *heterophylla*) –
a hybrid of *Quercus phellos* (willow oak) and
Quercus rubra (red oak) – was described in
1812 from a tree growing in Bartram's garden
that had first been discovered on an estate
nearby; its parentage was first remarked on
by Asa Gray (see pp104–109). This hybrid,
occurring elsewhere in the United States and
probably also in cultivation, is a medium to
large tree with leaf margins varying from
smooth to strongly toothed.

✤ In 1765, the same year that John Bartram
was appointed Royal Botanist for North
America by King George III, he and his son
William discovered *Franklinia alatamaha*
growing in 'two or three acres of ground
where it grows plentifully', according to William,
along the banks of the Altamaha River in
southeastern Georgia. On a return trip in

1773, William collected seed and brought it back
to John Bartram's garden where the tree was
successfully grown. Last seen in the wild in 1803,
all the plants in botanic gardens and arboreta
around the world are offspring of that seed. It
is a small, multi-stemmed deciduous tree and a
relative of the camellia, producing fragrant, single,
white waxy flowers with a central boss of golden
stamens. It flowers from midsummer to autumn,
when the leaves turn crimson. It prefers acidic,
well-drained soil and needs high summer
temperatures to grow well. Grow it up against
a sheltered sunny wall in cooler climates.

✤ *Kalmia* was named for Peter Kalm, who arrived
in Pennsylvania in 1748 and was befriended by
John Bartram. *Kalmia latifolia* (mountain laurel)
looks much like a rhododendron, apart from the
flowers. It likes acidic conditions and flowers in
midsummer. Clusters of flowers burst from buds
that look like icing sugar. In 'Alba', pale pink buds
open to white, pink-flushed flowers; 'Clementine
Churchill' has deep pink buds opening to bright
pink; 'Freckles' has pink buds and is cream with
purple spots and 'Ostbo Red' is red in bud,
opening to soft pale pink.

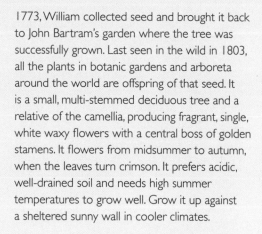

LEFT *Amelanchier bartramiana* (shadbush), named for
Bartram, grows in cold bogs in eastern Canada and
northeastern USA; it is a challenging garden plant.

TOP RIGHT After Bartram helped Peter Kalm on his
trip to North America, Kalm was commemorated in a
North American genus that includes *Kalmia latifolia*
(mountain laurel), noted for its pretty flowers.

RIGHT *Dionaea muscipula* (Venus fly trap), an
endangered species from North and South Carolina,
USA, was introduced into cultivation by Bartram.

Carl Linnaeus

DATE	1707–1778
ORIGIN	SWEDEN
MAJOR ACHIEVEMENT	LINNAEAN BINOMIAL SYSTEM

Linnaeus revolutionised natural history by introducing a system of classification based on the sexual organs of plants, simplifying their names into a concise binomial (two-word) description. Prior to this, plant descriptions had been long and complicated. French botanist Antoine de Jussieu, for example, had described coffee as *Jasminum, arabicum, lori folio, cujus semen apud nos coffee cicitur* – under Linnaeus's classification it was simply *Coffea arabica*. No wonder he said 'God created; Linnaeus organised' to describe his own contribution to science. Linnaeus named over 7,700 plants and 4,400 animals, and was the first to describe whales as mammals and to place humans with other primates.

Aloe succotrina,
fynbos aloe

Carl Linnaeus was born on 23 May 1707 in a turf-roofed homestead in Råsult, southern Sweden. He was the eldest of five children and the son of Rector Nils Linnaeus, a keen gardener and amateur botanist.

As soon as Carl became aware of his surroundings, his father decorated his cradle with flowers. Later, his father often carried the young boy to the garden or a meadow and set him on the ground with a flower to play with; even as a child Linnaeus wanted to learn plant names, and later played truant from school, preferring to search the nearby fields for plants.

At the age of 20, Linnaeus studied medicine, first at Lund University and later at Uppsala University, where Professor of Medicine Olof Rudbeck the Younger appointed him Demonstrator at the Botanic Garden. Employing a knowledgeable, enthusiastic second-year student instead of the usual jaded old lecturer caused a sensation, and classes of 70 to 80 suddenly became 400.

Around the same time, Linnaeus became aware of the limitations of existing systems of plant classification and decided to devise his own. Using the number and arrangements of the sexual organs, he grouped plants into class, order, genera and species, and introduced a binomial system for the genus and species. After revising an earlier thesis to incorporate this new system, he had this work presented to the Royal Society of Science by Rudbeck the Younger. Linnaeus wrote afterwards that the audience 'at first thought I was mad; but when I explained my intention they stopped laughing and promised their full support.'

A GROWING REPUTATION

In 1735, while on a trip to Holland, Linnaeus met a director of the Dutch East India Company, George Clifford, who kept a grand garden and zoo on his country estate. Linnaeus wrote that these were 'masterpieces of Nature aided by Art ... with shady walks ... artificial mounds and mazes', but it was the hothouses that excited him most of all: 'The first contained plants from southern Europe, the second treasures from Asia, like cloves, mangosteens and coconut palms ... the third plants from Africa including succulents like Aloes and *Stapelia* and the fourth, plants from the New World – cacti, passion flowers, calabash trees, bananas and camphor trees.'

Clifford was impressed by Linnaeus's ability to classify plants by opening a flower and examining its floral parts. When Clifford, a hypochondriac, suggested Linnaeus work as his physician and superintendent of his garden, Linnaeus accepted, receiving a salary of 1,000 florins a year, plus free board and lodging. In return, he supervised the hothouses, classified and catalogued Clifford's herbarium and garden plants, recorded every plant (such as *Aloe succotrina*, pictured on p56) in his *Hortus Cliffortianus* and monitored the health of his employer.

While in Clifford's employment, Linnaeus spent around three weeks in England, where he met Sir Hans Sloane, president of the Royal Society. Biographer John Lauris Blake noted in 1840 that Sloane 'did not pay [Linnaeus] that respect and attention that his merits deserved'; however, during these

weeks in England, Linnaeus did win over Philip Miller, superintendent of the Society of Apothecaries' garden at Chelsea, whose *Gardener's Dictionary* Linnaeus declared enthusiastically was 'not a dictionary for gardeners but for Botanists'.

Miller used current names, such as *Symphytum consolida major, flore luteo*, Linnaeus later recalled. 'I remained silent, with the result that he said the next day: "This botanist of Clifford's doesn't know a single plant." When he again began using them I said "Don't use such names, we have shorter and surer ones" then gave him examples.' Linnaeus also travelled to Oxford to meet Sherardian Professor of Botany, Johann Jacob Dillenius, who disagreed with him, too. However, after discussing what Dillenius had marked as errors in the soon-to-be published *Genera Plantarum*, Linnaeus dissected some of the flowers, showed him the individual parts and proved that his system was correct.

Coffea arabica: Linnaeus reduced de Jussieu's Latin description from ten words to just two.

Soon, Linnaeus was fêted everywhere except Sweden, where his work was ridiculed. 'I was the laughing stock of everyone on account of my botany', he wrote later. However, his fortunes gradually changed and on 27 October 1741 he became Professor of Botany at Uppsala University, inheriting its run-down botanical garden. In 1685, it had contained 1,800 plants; by 1739, there were only 300. The Senate financed the construction of a glasshouse and extended the garden, the Prince Regent sent him a racoon to add to the menagerie and demanded he examine it, and goldfish were sent from England. Linnaeus wrote: 'Do send the goldfish tomorrow, so that I can see what I have wanted all my life but never dared to hope for.' There was also a talking parrot. If Linnaeus was late for his lunch it would squawk 'Twelve o'clock, Mr Carl!' until he appeared.

LINNAEUS'S LEGACY

Linnaeus wrote papers on lemmings and ants, a phosphorescent Chinese grasshopper, Siberian buckwheat, rhubarb, fossils and crystals, and he investigated leprosy and epilepsy – amounting to 170 papers and dissertations in all. His rules for plant classification were elucidated in *Fundamenta Botanica* (1736), expanded in his *Classes Plantarum* (1738) and reached their zenith in his two volumes of *Species Plantarum* (1753) – a work he called 'the greatest in botany'. The 1,200 pages cover 5,900 species in 1,098 genera and are the basis of modern taxonomy. He also wrote *Systema Naturae*, its zoological equivalent.

Linnaeus's British contacts were among the earliest to endorse his work. Among them were naturalist and apothecary John Hill, and Philip Miller, who incorporated Linnaeus's names and taxonomy into the eighth edition of his *Gardener's Dictionary*. Sir Joseph Banks took Daniel Solander, Linnaeus's pupil, on James Cook's first voyage then employed him as curator, and James Edward Smith (see pp74–79) bought Linnaeus's collections after his death and established the Linnean Society of London – thereby securing the future of Linnaeus's binomial system. This system was a work of powerful simplicity, and we depend on its genius today.

Carl Linnaeus:
INSPIRATION FOR GARDENERS

✤ Like many of his predecessors, Linnaeus stuck to Latin for his binomial system. Being a historical language, Latin wasn't likely to evolve, and it has now become a global lingua franca for botanists.

✤ In Linnaeus's system of plant names, genus comes first, followed by any individual or specific names. Genus: *Magnolia*; species: *grandiflora*. These can then be divided further, into subspecies (subsp.), variety (var.) and form (f.).

BELOW *Magnolia grandiflora* (southern magnolia) from the USA; Linnaeus's system tells us that this is the *grandiflora* species within the *Magnolia* genus.

✤ Cultivars (cultivated varieties of plants) were a twentieth-century addition to Linnaeus's system, and can be found in single quote marks in plant names, as in *Camellia japonica* 'Apollo'. To create a cultivar, a plant with a particular combination of features is selected and then vegetatively propagated (a form of asexual reproduction, requiring only one parent) to produce an identical version of the parent plant. There can also be groups of cultivars with similar features; the *Acer palmatum* Dissectum Group all have the fern-like leaves of the *Acer palmatum* 'Dissectum' cultivar.

✤ There are usually nuggets of interesting or useful information in a plant's Latin name. For example, *Paulownia tomentosa* took its name from the nineteenth-century Russian princess Anna Paulowna and the Latin for

BELOW *Camellia japonica* (Japanese camellia), one parent of the famous *Camellia* × *williamsii* hybrids. The '×' in a plant name always denotes a hybrid.

'woolly with matted hair' – the latter being a description of its leaves. Linnaeus loved the honeysuckle he named *Linnaea borealis*, describing it as 'a plant from Lapland, lowly, insignificant, disregarded, flowering but for a brief space, from Linnaeus who resembles it'; *borealis* means 'northern'.

✤ The 'x' in a plant's name means it is a hybrid. A group of hybrids from the same parentage are often given a collective name, such as *Camellia* × *williamsii* (a cross of *Camellia japonica* and *Camellia saluenensis*). Hybrids bred from two different genera are written like × *Mahoberberis* (a cross of *Mahonia* and *Berberis*). However, there are very few intergeneric hybrids in cultivation.

✤ A new variety of plant may also be sold under its trade name, which protects the breeder's rights over its propagation – for instance, Magnolia 'Shirazz'™ ('Vulden').

TOP *Paulownia tomentosa* (empress tree) was named for Russian princess Anna Paulowna and *tomentosus*, the Latin for woolly and matted, on account of its hairy leaves.

ABOVE The names of intergeneric hybrids – crosses from two different genera – start with an '×', such as this × *Mahoberberis*.

Jeanne Baret

DATE	1740–1807
ORIGIN	FRANCE
MAJOR ACHIEVEMENT	BOTANIST AND PLANT COLLECTOR

When Philibert Commerçon joined Louis-Antoine de Bougainville's 1766 expedition to circumnavigate the world, he took Jeanne Baret with him as his valet and botanical assistant – disguised as a boy. An expert botanist due to her knowledge of medicinal plants, Baret thereby became the first woman to circumnavigate the globe, and proved herself a resilient plant collector. Among her discoveries were the first specimens of the famous *Bougainvillea spectabilis* from the forests of Brazil. As expedition naturalist, Commerçon received considerable recognition for his work; Baret has only recently been commemorated through the name of a plant.

Ixora coccinea,
burning love

Sometime between 1760 and 1764, Jeanne Baret, a country girl as well as a knowledgeable botanist, became the housekeeper and lover of naturalist and botanist Philibert Commerçon, friend of Voltaire and correspondent of Linnaeus (pp56–61). Several years later, in 1766, Louis-Antoine de Bougainville was selected to command an expedition to discover flora and fauna of commercial value that would flourish in France or her colonies, and Commerçon was invited to join it. It was the first expedition to include a professional naturalist paid for by a sovereign, and aimed to be the first French circumnavigation of the world. The two ships, *Boudeuse* and *Étoile*, were to sail from France to Montevideo, Rio de Janeiro, Buenos Aires, the Falkland Islands, Tierra del Fuego, the coasts of Magellan and the South Pacific Islands, returning via Australia, the Molluccas, Java and Réunion; Commerçon helped to plan the itinerary.

Commerçon's senior role permitted him a servant, but the French Navy did not allow women on board ships, so in a daring plan, for which she became famous, Jeanne bound her breasts in tight strips of linen, wore baggy clothes, changed her name to Jean, and joined Commerçon as his botanical assistant and personal valet. Although there were mutterings among the crew, her plan succeeded until 1767 when the ships reached Tahiti and the indigenous people saw through her disguise. Nothing is reported of their relationship after her discovery, other than that she continued to be dedicated to him and to botanising in search of new plants.

AN INTREPID PLANT HUNTER

In *A Voyage Around the World*, his published account of the trip, Bougainville recalled: 'For some time there was a report in both ships, that the servant of M. de Commerçon, named Baré, was a woman. His shape, voice, beardless chin, and scrupulous attention of not changing his linen, or making the natural discharges in the presence of any one, besides several other signs, had given rise to, and kept up this suspicion. But how was it possible to discover the woman in the indefatigable Baré, who was already an expert botanist, had followed his master in all his botanical walks, amidst the snows and frozen mountains of the straits of Magalhaens, and had even on such troublesome excursions carried provisions, arms, and herbals, with so much courage and strength, that the naturalist had called him his "beast of burden"?' Commerçon was equally complimentary, writing that she traversed the highest mountains with agility and the deepest forests without complaint: 'Armed with a bow like Diana ... armed with intelligence and seriousness like Minerva ... she eluded the snare of animals and men, not without many times risking her life and her honor.' Commerçon praised her for the numerous plants she collected, the herbaria she constructed and the collections of insects and shells she curated.

Most populist literature focuses on Jeanne Baret's story of deception, rarely mentioning her botanical skills. Yet she and Commerçon worked together, and when he suffered from ill health, as he often did on the voyage, she collected alone. Such a collaborative partnership in the discovery of botanical and

zoological wonders is unique in the history of science. Every time the ship docked, the pair disembarked to collect samples. They recorded their observations and were amazed at the wealth of diversity they encountered. Commerçon wrote of Madagascar: 'I can announce to naturalists that this is the true Promised Land. Here nature created a special sanctuary where she seems to have withdrawn to experiment with designs different from those used anywhere else. At every step one finds more remarkable and marvellous forms of life.'

Bougainvillea spectabilis, a plant believed to have been discovered by Jeanne Baret in Rio de Janeiro in 1767.

One of the expedition's most notable collections is thought to have been made by Jean Baret (if not, she was certainly with Commerçon when the collection was made). On the basis of notes made by François Vivez, the *Étoile*'s surgeon, the hypothesis has been put forward that because of the varicose ulcers and gangrene that constantly afflicted him, Commerçon was possibly too ill to collect plants during his stay in Rio de Janeiro between 13 and 15 June 1767. All of the plants from that location were therefore gathered by Baret – one of many times she might have assumed the role of chief botanist. It was here that a flamboyant scrambling plant was collected and later named *Bougainvillea spectabilis* in honour of the expedition's commander. Five herbarium specimens exist from the voyage, three with the 'Buginvillaea' written on them, in what appears to be Commerçon's handwriting; the plant was officially named by Antoine Laurent de Jussieu in 1789, based on these specimens and notes.

A WELL-DESERVED TRIBUTE

Commerçon and Baret (though uncredited) amassed 1,735 specimens for the French National Herbarium at the National Museum of Natural History in Paris. Of these, 234 were from Madagascar, 144 from Mauritius, 84 from Brazil and 50 from Java; many formed the basis of new species. They also documented unknown plants introduced to new habitats by human populations, such as *Ixora coccinea* (pictured on p62), which was brought to Tahiti by early Polynesian settlers.

When the ship stopped in Mauritius, Commerçon and Baret disembarked, leaving the rest of the expedition to visit Madagascar and Réunion. When Commerçon died, aged 46, after years of ill health, Baret married French officer Jean Dubernat on Mauritius, returning to France several years later in 1774, and becoming the first woman to sail across the Pacific and to circumnavigate the world.

Over 70 different types of flora and fauna were named after Commerçon, yet despite her vital role in the expedition, not a single species was named after Baret. Commerçon had proposed *Baretia* for a Madagascan tree but that name had already been taken. It was not until 2012 that a new species of *Solanum* from the Andean cloud forest of Ecuador and Peru was named *Solanum baretiae*. A distinguishing feature of the plant is its numerous different leaf shapes – simple or divided, with varying numbers of leaflets, the perfect plant to commemorate Jeanne Baret, botanist and master of disguise!

Jeanne Baret:
INSPIRATION FOR GARDENERS

❖ Bougainvillea, along with hibiscus, are quintessential exotic garden plants. A plant in full bloom against a bright blue sky is a sight never to be forgotten. The colour does not come from the flowers, which are small and white, but the brightly coloured bracts (modified leaves) surrounding them. They can be planted in a conservatory in cooler climates, but as pot plants, can be put outside on hot summer days. They need bright light, blooming prolifically after a cool winter rest,

BELOW Baret made important botanical discoveries; bougainvilleas are popular ornamental plants today, with tiny white flowers, surrounded by vivid bracts.

but when temperatures are too low they take a long time to recover in spring. Keep the compost just moist in winter, reducing watering further if the leaves fall, then increase watering as growth increases in spring.

❖ For an eyecatching display, allow as many main stems to grow as possible, as cutting them back to the chosen height encourages flowering side shoots. Cut side shoots back to two buds in late winter to keep plants under control and encourage flowering. Feeding with high-potash fertiliser also encourages flowering in cooler climates. In tropical or Mediterranean climates, these plants are

excellent on large pergolas or draped over boundary walls where their thick growth and spiny stems provide a natural deterrent to intruders.

✤ Mauritius, where Commerçon and Baret disembarked, is home to some extraordinary native plants and pollinators, many now endangered. As one of the few places in the world where there were no honeybees and few common tropical pollinators, many native plants were pollinated by geckos, attracted to the flowers' coloured nectar. *Nesocodon mauritianus*, found on Mauritius and nowhere else in the world, has pale blue flowers and its blood red nectar is produced in such volumes that it drips from the flowers. *Roussea simplex,* native to the mountain forests, with its orange-yellow nectar is a botanical oddity. It is pollinated and seeds are dispersed by the Mauritian blue-tailed day gecko (*Phelsuma cepediana*), the only known plant in the world to have the same pollinator and seed dispersal agent.

TOP *Geum magellanicum* was collected by Commerçon in 1767; one of the original specimens can be found at the University of Montpellier's herbarium in France.

ABOVE New species are still being discovered on Mauritius, centuries after Baret's voyage. The endangered *Nesocodon mauritianus* was found in 1976, growing in a single location, on a near-vertical cliff.

Joseph Banks

DATE	1743–1820
ORIGIN	ENGLAND
MAJOR ACHIEVEMENT	JAMES COOK'S BOTANIST ON THE *ENDEAVOUR*

Joseph Banks gained fame as a naturalist and botanist after James Cook's first voyage. He also promoted other people's research, while his own work developing the collections at the Royal Botanic Gardens at Kew transformed it into the great garden it is today. Banks's knowledge was respected by contemporaries such as Johann Reinhold Forster, naturalist on Cook's second voyage, who stated upon the death of Carl Linnaeus (see pp56–61): 'great as the loss of Linnaeus must certainly be to science, it will not be severely felt while we have so enlightened botanists as Mr Banks and Dr Solander.'

Fuchsia coccinea,
scarlet fuchsia

As a boy, Banks explored the Lincolnshire countryside and developed a keen interest in natural history. He was educated at Harrow and Eton, botanised in England and Wales, created his own herbarium and studied Linnaean taxonomy. When he went to Christchurch College, Oxford, the Professor of Botany Humphry Sibthorpe, who had not published a scientific paper and delivered just one public lecture during his 35-year tenure, did not object when Banks hired a lecturer of his own. Banks left Oxford without a degree, however, and immersed himself in London society. He then secured a position onboard the HMS *Niger* heading to Newfoundland on fishery protection duties; while he was there, Banks collected birds, insects and rocks, and observed the habitats of Inuits. His botanical collections and notes confirmed his proficiency in botany and he was elected Fellow of the Royal Society, aged 23. The success of this voyage whetted Banks's appetite to see the world. When the Government and Royal Society decided to send a ship to observe the transit of Venus in Tahiti in 1766, the captain, James Cook, also had secret instructions to discover Terra Australis. The Royal Society asked if Banks could join HMS *Endeavour* for 'the advancement of useful knowledge'; Banks, a man of considerable means after the death of his father in 1761, paid his own passage plus that of his entourage of eight, including the Swedish botanist Daniel Solander, and Sydney Parkinson, a scientific artist.

COOK'S FIRST VOYAGE

The *Endeavour* left Plymouth on 25 August 1768, bound for South America, with Banks and his staff collecting plants in Madeira, Rio de Janeiro and Tierra del Fuego. On leaving Tahiti, the *Endeavour* explored the coasts of New Zealand and Australia before returning home in July 1771. Before his death at sea from malaria and dysentery, Parkinson produced 280 paintings and 900 sketches and drawings. On his return to England, Banks commissioned over 700 copper engravings, ready for publication, but his *Florilegium* was not published until two centuries later. Overall, Banks collected 3,600 species of dried plants, of which 1,400 were new to science, plus seeds, plants and drawings. He was fêted by society, had an audience with George III at St James's Palace, and instantly became a highly respected botanist and naturalist.

Within a few months, a second voyage was planned. Banks automatically assumed he would be included, and this time his party was to include botanist Solander, four artists, two secretaries, six servants and two horn players. When Cook complained, Banks petulantly withdrew and went to Iceland, giving some of the ship's ballast of lava to the Chelsea Physic Garden for their rock garden and to Kew, where it provided the structure for the Moss Garden, on his return.

KEW GARDENS

Banks was well aware of Princess Augusta's exotic or physic garden at Kew; the *Westminster Journal* for 24–31 August 1771 reported, 'Dr Solander and Mr Banks had the honour of frequently waiting on his Majesty in Richmond, who is in a

course of examining their whole collection of drawings of plants and views of the country'. Banks became a considerable horticultural and botanical influence on George III, with William Townsend Aiton, who had been responsible for the small physic garden since 1759, admitting, 'this establishment is placed under the direction of Sir Joseph Banks'. They regularly met on Saturdays, one equerry noting: 'the King continued his walk with Sir Joseph Banks about 3 hours, they first visited the exotic garden, then walked through the Richmond gardens.'

Gradually, Banks transformed Kew into a centre for global transfer of plants, insisting that 'as many new plants as possible should make their first appearance at the Royal Gardens'. Banks revealed: 'I have long ago given up the collection of plants in order that I might be better able to promote the King's wishes to make the Royal Botanic Gardens at Kew as respectable as possible, so that everything that comes to me from whatever quarter it may be, is instantly sent to Kew.' Banks was ably supported in these ambitions by William Aiton, and the swift, seamless conversion of Kew from royal pleasure grounds to botanic garden, owed much to their amicable working relationship.

In 1773 nearly 800 species were planted, most from North America, and Banks began to send out plant hunters. In 1774, Francis Masson, Kew's first collector, returned with 80 different species from the Cape of Good Hope, including *Protea*, *Crassula*, *Mesembryanthemum*, *Erica*, *Pelargonium*, *Ixia* and *Gladiolus*, and a special 'Cape House' was built to house his introductions. Masson believed that he added 400 species to the garden. Banks was delighted, believing these collections enhanced Kew's 'superiority which it now holds over every similar Establishment in Europe'.

This plate depicting *Banksia serrata* (saw banksia), from Banks's *Florilegium*, illustrates the level of artistic skill needed to complete this drawing.

Banks also reminded government officials, military officers, merchants and missionaries to consider Kew's needs during their travels. Sir John Murray, on active service in India with the British Army, despatched several packages of seed; the Governor of St Helena sent plants; and the Receiver General made large donations. When a precious specimen of *Fuchsia coccinea* (scarlet fuchsia, pictured on p68) from South Africa was presented to Kew in 1778, Sir Joseph Banks carried it into the garden on his head, not trusting anyone else to transport it. Sir George Staunton sent Banks a nutmeg tree, some mangosteens and six pots of plants. Bulbs arrived from Adam Afzelius, Swedish botanist to the Sierra Leone Company, and Governor Phillip returned from Australia with over 82 tubs and boxes of plants. Marquess of Lothian, William Kerr introduced many plants, too, including *Rosa banksiae*. A delivery in 1793 of nearly 800 pots, the largest single despatch ever made to Kew, filled HMS *Providence* when it docked at Deptford on the River Thames after Bligh's second voyage to source breadfruit. During the reign of George III some 7,000 new species were introduced from abroad, and the botanical mastermind behind it all was Sir Joseph Banks.

Joseph Banks:
INSPIRATION FOR GARDENERS

✤ *Rosa banksiae* (Banksian rose) is a slender climber with few spines on the stems. It likes warmth and sunshine and needs a position against a sheltered sunny wall in cooler climates. There are several desirable cultivars: 'Alba Plena' the white, double-flowered, violet-scented form from which the species was described, was introduced by Kew collector William Kerr from Canton in 1807. 'Lutea', the most familiar form in gardens due to it being the most hardy and floriferous, has dainty double yellow flowers and little scent; it was introduced through the RHS from China in 1824. 'Lutescens' has yellow, fragrant flowers, while var. *normalis* has white, fragrant flowers. The flowers are not borne on laterals from the previous year's growths, as in most ramblers, but on twigs produced by the

laterals, so stems will be two or three years old before they flower. Don't prune annually, just remove older stems periodically, once flowering is over.

✤ *Banksia* was first collected by Joseph Banks and Daniel Solander, at what was later named Botany Bay, due to the sheer quantity of plants Banks and Solander found there. There were four species in this first collection: *B. serrata* (saw banksia), *B. integrifolia* (coast banksia), *B. ericifolia* (heath-leaved banksia) and *B. robur* (swamp banksia). *Banksia* plants must have well-drained soil to survive. Fork finely chopped forest bark or gravel into heavier soils, or plant on a mound and mulch with gravel. They rarely need feeding but if you do use fertiliser, make sure it is phosphate-free; phosphates can kill the plants.

✤ Francis Masson, sent out on expedition by Banks and Kew, introduced various new pelargoniums in 1774, including *P. cordifolium*. This has pale pink flowers with dark veins, and large, grey-green tri-lobed leaves, with the underside covered with silvery hairs. It is also lemon-scented. *P. grandiflorum* was introduced by Masson after his second expedition to South Africa in 1797.

LEFT Banks sent Francis Masson to South Africa to search for plants; one of the treasures he returned with was *Pelargonium grandiflorum*.

RIGHT *Rosa banksiae* 'Lutea' (yellow Banksian rose, formerly Lady Banks rose), named for Banks's wife. It needs a sheltered, sunny spot in a mild climate or conservatory, protected from winds and the cold.

James Edward Smith

DATE	1759–1828
ORIGIN	ENGLAND
MAJOR ACHIEVEMENT	FOUNDER OF THE LINNEAN SOCIETY

Passionate about plants and botany from childhood, James Edward Smith's enlightenment came on 10 January 1778, the day Carl Linnaeus died (see pp56–61). Despite it being midwinter, the 18-year-old was deep in the ditches of Norwich Castle collecting his first plants, having bought his first botanical book the previous day – John Berkenhout's *Outlines of the Natural History of Great Britain and Ireland*, arranged according to the new Linnaean principles. Later in life, Smith would purchase the Linnaean Collections and establish the Linnean Society of London, elevating the status of British science and popularising botany.

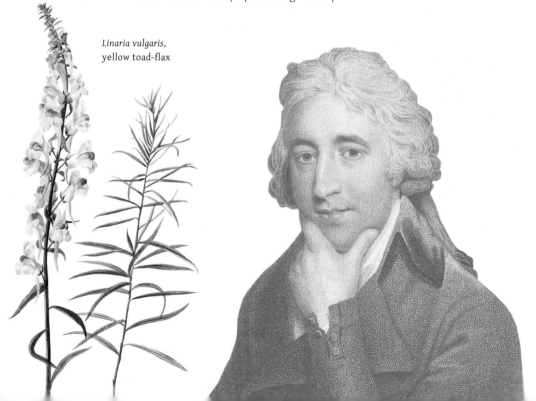

Linaria vulgaris,
yellow toad-flax

James Edward Smith was born on 2 December 1759 at 37 Gentleman's Way, Norwich, the eldest of eight children of a wealthy textile merchant. A delicate, indulged favourite, he was home-schooled – one of his earliest memories was 'tugging ineffectually with all my infant strength at the tough stalks of the wild zuchery [chicory] on the chalky hillocks about Norwich'.

Destined for the silk trade, Smith was determined to study botany instead, but as there were no professional positions available in this field, he chose to train as a physic, where a knowledge of plants would be essential. In 1781 he began studying medicine at the University of Edinburgh, where his botany lecturer was Dr John Hope, one of the earliest teachers of the Linnaean system. Here Smith gained a reputation for both fastidious attention to botanical detail, and social skills, forming a natural history society for the students of the university. Two years later, Smith moved to London and St Bartholomew's Hospital Anatomy School (where he was famously squeamish), carrying an introduction from Hope to Sir Joseph Banks, then President of the Royal Society.

THE LINNEAN SOCIETY

On 23 December 1873, the 24-year-old Smith attended a breakfast party, at which Joseph Banks was also present. Banks had received a letter from Sweden offering him the entire collection of Linnaeus's books and specimens for the sum of 1,000 guineas. Banks was forced to turn down the offer due to his financial situation, but he suggested that Smith purchase it instead. Smith responded quickly; aided by a loan from his father, he secured a deal before the rest of the world had caught on – including the king of Sweden (who was in Italy at the time) – and outwitting potential bidders Catherine the Great of Russia and John Sibthorp, Sherardian Professor of Botany at Oxford.

The vast collection arrived in London on 10 October 1784 and Smith, who was most interested in the 14,000 plant specimens, suddenly became a patron of learning and supporter of British advancement. As one friend wrote, 'the nobel purchase of the Linnaean cabinet, most decidedly sets Britain above all other nations in the botanical empire'. The following winter, Smith catalogued the collections with the help of Banks and his librarian, Jonas Dryander, and a friend, John Pitchford, urged Smith to prepare 'a Flora Britannica, the most correct that can appear in the Linnaean dress'. Smith was soon elected a Fellow of the Royal Society, and in June 1786, he set out on his Grand Tour, visiting herbaria and botanical libraries en route. At Versailles, he even encountered Louix XVI, noting in a letter home that the monarch had 'grown fat and had an agreeable countenance, his ears are remarkably large and ugly, he had some very fine diamonds in his hat'.

In 1788 Smith moved to Great Marlborough Street in London, and on 18 March he founded the Linnean Society of London with the botanist and Bishop of Carlisle, Samuel Goodenough, and the entomologist Thomas Marsham. At the first meeting, on 8 April 1788, Smith was elected president, and delivered an

introductory talk on the rise and progress of natural history. Marsham became secretary, Goodenough was made treasurer, and Dryander librarian. Banks joined as an honorary member. Smith's house quickly became a hub of scientific activity.

Smith soon came to the attention of Queen Charlotte, wife of George III of England and a keen amateur botanist, and in October 1792 Smith was invited to teach botany and zoology to the Queen and Princesses at the royal residence, Frogmore House. However, the account of his recent journey, *Sketch of a Tour on the Continent*, contained passages praising Rousseau and referring to Marie Antoinette as Messalina (the ruthlessly ambitious, promiscuous third wife of the Roman emperor Claudius), which outraged the Queen, and the offer was subsequently withdrawn.

LORD TREASURER OF BOTANY

On 1 March 1796 Smith married Pleasance Reeve, daughter of a Lowestoft attorney and merchant, whom he described as a 'very fine person, with an angelic countenance and a very pretty fortune' (she was also great aunt of Alice Pleasance Liddell, immortalised in *Alice in Wonderland*). Soon after, Smith retired to Norwich, visiting London for two or three months each year to deliver a series of lectures at the Royal Institution – spending the rest of his time in his museum, with his old-fashioned cabinets from Uppsala looking very out of place (it is said in an 1829 issue of *The Philosophical Magazine* that 'the relics of Mohammed are not enshrined with more devotion'). Every day, he wrote from 9am until 3pm, then from 7 to 9pm, also replying to numerous correspondents. He was re-elected president of the Linnean Society annually until his death.

Cichorium intybus (chicory) was one of the plants that Smith later recalled learning about as a young child.

After completing his three-volume *Flora Britannica* (cataloguing British plants, such as *Linaria vulgaris*, pictured on p74), Smith was chosen by John Sibthorp's executors to edit *Flora Graeca*. In 1807 he published his most successful work, *The Introduction to Physiological and Systematic Botany*. Five editions were produced during his lifetime, and two posthumously. He also contributed 3,348 botanical articles and 57 biographies of botanists to Rees's *Cyclopaedia* – including one on William Curtis. In 1814, the Prince Regent became Patron of the Linnean Society, and Smith was knighted for his services to science.

In the last seven years of his life, Smith worked on *The English Flora*, often described as his 'last and best work', with descriptions drawn from years of critical study. The first two volumes appeared in 1824, a third in 1825, the fourth in 1828, days before his death; the fifth, with a contribution on mosses by Sir W. J. Hooker, appeared between 1833 and 1836. Smith died in Norwich, on 17 March 1828, and was buried at St Margaret's Church, Lowestoft, in the Reeve family vault. Josef August Schultes, Professor of Natural History at the University of Landshut in Bavaria, perhaps best summarised Smith's achievements, hailing him during his lifetime as 'the only orthodox Botanist in Europe' and 'Lord Treasurer of Botany'.

James Edward Smith:
INSPIRATION FOR GARDENERS

✤ In the 36-volume *English Botany* – published between 1791 and 1814 by James Edward Smith and the publisher and illustrator James Sowerby – Smith remarked that the best place to find *Vinca minor* (lesser periwinkle) growing wild was at Honingham Church, Norfolk. *Vinca minor* is useful groundcover in most soils, thriving in shade or light, dry shade under trees. If the foliage becomes untidy, cut it hard back at the end of winter. There are several garden-worthy varieties: *Vinca minor* 'La Grave', an evergreen with glossy oval leaves and bright blue, rounded flowers, was found by the horticulturalist E. A. Bowles in a churchyard in La Grave, France; *Vinca minor* f. *alba* 'Gertrude

BELOW Boldly coloured *Dryandra formosa* (showy dryandra), also known as *Banksia formosa*, was named for Banks by botanist Carl Linnaeus's son.

BELOW Smith deemed a local churchyard the best spot for *Vinca minor* (lesser periwinkle). Try *Vinca minor* 'Atropurpurea' (dark purple-flowered periwinkle) in shady spots.

Jekyll' has white flowers and small, shiny green leaves; and *Vinca minor* 'Atropurpurea' bears small plum-coloured flowers. They can become a weedy species in borders unless carefully controlled, and are better naturalised in a wild area where they will happily spread.

✤ Chicory, which Smith found 'in the chalky hillocks about Norwich', is grown in the vegetable garden as 'Whitloof' or 'Belgian' chicory. The plants are blanched before harvesting to remove bitterness. Chicory can be forced – lifting the roots prematurely and moving them to a dark place to encourage premature growth. It needs an open sunny site. Sow in spring, keeping the plants watered and weed-free for lifting from late autumn until midwinter. The ideal size for forcing is between 3.5 to 5cm (1¹/₂–2in) diameter across the top. Trim the leaves to 2.5cm (1in) above the neck, then cover with straw outside or indoors, in flat trays of sand. Force a few at a time in 23cm (9in) pots containing moist compost, cutting the roots to length. Exclude light by covering with a similar-sized pot with the drainage holes covered, or a large box to discourage humidity. Maintain temperatures of 10°C to 18°C (50–64°F). Blanching takes at least three weeks.

TOP Smith used *Ilex aquifolium* (common holly) to comment on the natural classification of plants in his notes, which were published posthumously in 1832.

ABOVE *Carex riparia* (greater pond sedge), one of many plants illustrated in Sowerby's and Smith's *English Botany*. It is an ideal plant for a bog garden.

79

Philipp Franz von Siebold

DATE 1796–1866	
ORIGIN GERMANY	
MAJOR ACHIEVEMENT *FLORA JAPONICA*	

Philipp Franz von Siebold, a German physician, worked for the Dutch government and introduced Western medicine into Japan. At a time when the country was closed to the rest of the world, he practised medicine and collected plants and Japanese artefacts. He was expelled from the country twice, the first time after being accused as a Russian spy. His daughter Kusumoto Ine became the first female doctor in Japan and Siebold introduced over 800 species of plants into Europe, including hostas, hydrangeas, *Cercidiphyllum japonicum*, *Chaenomeles japonica* (pictured below) and *Rosa rugosa* 'Alba'.

Chaenomeles japonica,
Japanese quince

Philipp Franz von Siebold was born into a well-connected family of physicians in Würzburg, Germany. After university, he went into the family profession, specialising in ophthalmology. After the death of his father, relatives wrote to one of the many influential family friends, Franz Joseph Harbaur, Inspector-General of the Dutch Army and Navy medical corps, to enquire about a position for him; around about the same time, Siebold developed a love of natural history.

In 1822 Siebold was appointed 'Surgeon First Class' in the Dutch East-Indies Army for the Dutch Government Service and headed to their colonies (in what is now Indonesia), leaving Rotterdam on 23 September 1822. On the long journey at sea, Siebold learned to speak basic Dutch and Malay. Arriving in Jakarta, a further post was added when he was appointed 'Surgeon First Class in charge of research into natural science in the Japanese Empire' at the Dutch trading settlement of Dejima in Japan, and told to study not just local natural history but also the land and its peoples, to increase opportunities for trade with that country.

LIFE IN JAPAN

Siebold arrived in Dejima on 12 August 1823, having survived a raging storm. He remained there for six years, collecting wild plants in the hilly, densely wooded countryside around Nagasaki and gaining respect, friends and admirers for his use of belladonna (*Atropa belladonna*) in treating cataracts, and introducing Western medicine to Japan. Siebold collected plants at every opportunity, particularly while visiting the local medical school that he established. He also began a small botanic garden behind his home and many of the plants were planted there, then shipped to the Netherlands. Siebold also bought plants from local growers outside the city, and encouraged colleagues and assistants to send him plants, too. By 1825 he had assembled more than 1,000 plants in his garden, adding to his collections by creating another garden around his medical school.

During the seventeenth and eighteenth centuries, Dutch merchants made an annual pilgrimage to pay homage to the ruling shogun in the city of Edo (now Tokyo). On one such occasion Siebold was introduced to several important Japanese natural scientists, including the artist Katsuragawa Hoken, known as Wilhelm Botanicus, a physician to the shogun and later one of Siebold's most talented botanical artists, and the botanist Keisuke Ito, who presented Siebold with a herbarium and collected with him in the hills around Nagasaki.

In 1827, the Dutch government recalled Siebold to the Netherlands, to devote himself to processing his collections, but as his ship was preparing to sail with his plants, artefacts and hand-drawn maps, a huge storm struck Nagasaki, rendering the ship unseaworthy. An undercover agent had already reported that Siebold was leaving with forbidden maps and a kimono jacket bearing the shogun's emblem – three golden leaves of *Asarum nipponicum*; the former was strictly forbidden, and the latter was seen as a grave insult. Siebold was placed under house arrest, tried,

revealed to be German not Dutch, accused of spying for the Russians, then ignominiously banished from the country. (While under house arrest, however, he managed to grow seeds taken from the fodder for his goat and made herbarium specimens of the plants.)

RETURN TO EUROPE

Siebold's arrival in Europe coincided with Belgium's battle for independence from the Netherlands so he rushed to remove the manuscripts, specimens and collections he had sent to Antwerp, Brussels and Ghent to the safety of Leiden. The collection of 80 of his most precious Japanese plants remained in Ghent Botanical Garden and inaccessible to him for several years. However, following a string of angry complaints about the loss of his plants and the fact that they had been made available to Ghent nurserymen who were, in turn, selling them for profit, Siebold was donated one plant of each by the director of the gardens as a gesture of goodwill.

Siebold settled in Leiden, curated his collections and began writing. His three works revealed Japan to the West: *Nippon* (1832–58), *Fauna Japonica* (1833–50) and *Flora Japonica* (1835–42), with the German botanist Joseph Gerhard Zuccarini, who provided botanical descriptions for the first section. Siebold and his assistant contributed plant specimens, information on localities, Japanese and Chinese vernacular and scientific names, and French plant descriptions.

Siebold was particularly interested in economic plants, growing tea from Japan, China and Assam; trialling the edible root *gobo*, or burdock (*Arctium lappa*); and the Chinese weeping white mulberry (*Morus alba*). However, his most significant crop was sweet potato (*Ipomoea batatas*), which had been introduced to Japan in the early eighteenth century. He grew this for four years in his 'Jardin d'Acclimatation' in Leiden, listing four different cultivars in the catalogue of his nursery, Von Siebold & Comp., in 1856.

Siebold gained fame for treating cataracts with *Atropa belladonna* (deadly nightshade) and bringing Western medicine to Japan.

In 1859, when he was officially pardoned by the Japanese and the country had opened up to the West, Siebold returned to Japan, where he wrote a catalogue of 263 plant species that he had collected. He collected large numbers of ornamental plants for his nursery, dispatching crates to Leiden on 7 January 1861, and again three months later. He also made several botanical trips into the hills above Yokohama in April and his assistant copied various documents for him, again landing him in trouble; he was asked to leave for a second time on 31 October 1861.

Back in Europe, Siebold oversaw the cultivation of 280 new introductions in his garden and held plant sales in 1865 and 1866, alongside an exhibition of 10,000 new plants from Japan, of 800 species and varieties. Even in his later years, Siebold continued working energetically until he contracted typhoid from contaminated water and died in Munich on 18 October 1866. Twelve of his plants are still growing in the University of Leiden Botanic Garden – among them, *Clethra barbinervis* (Japanese clethra), *Parthenocissus tricuspidata* (Boston ivy) and *Zelkova serrata* (keaki).

Philipp Franz von Siebold:
INSPIRATION FOR GARDENERS

❖ *Fallopia japonica* (Japanese knotweed), introduced to Europe by Philipp Franz von Siebold around 1829, illustrates the dangers that can be posed by the introduction of garden plants. The original plant was propagated in Holland and marketed around Europe by Siebold's Leiden nursery. It was sold primarily as an ornamental plant. Siebold brought back just one female plant and all Japanese knotweed in Europe is a clone of that original plant. If you have this on your property, ensure it is removed immediately by specialist contractors.

❖ One of the plants first described in 1839 in *Flora Japonica* was from a collection made by Siebold in the village of Kosedo, on Kyushu Island, near Nagasaki. This was *Wisteria brachybotrys* (silky wisteria). It had been grown for centuries in Japan for its sumptuous scent. It has large individual flowers, and stems and leaves are covered in silky golden hairs when they first emerge.

BELOW One of Siebold's introductions to Europe, *Fallopia japonica* (Japanese knotweed), is hugely invasive and has wreaked widespread havoc.

BELOW RIGHT *Wisteria brachybotrys* f. *albiflora* 'Shiro-kapitan' (silky wisteria); this species was first described in 1839 by Siebold and Joseph Zuccarini, based on a Siebold collection.

This plant was listed in Siebold's catalogue of Japanese plants offered by his nursery. The first record in England appears to have been in the 1881–82 catalogue of the Veitch & Sons nursery in London's Chelsea. There are several commonly available cultivars: *Wisteria brachybotrys* f. *albiflora* 'Shiro-kapitan', with white tassels of flowers up to 15cm (6in) long; 'Okayama', a pale lilac-flowered cultivar; 'Shiro-beni', with grey-pink young buds and bracts, and white flowers; and 'Showa-beni', which has pink flowers.

❧ Siebold introduced over 20 hosta species and forms, in those days classified in the genus *Funkia*. Two species were named for him, *Hosta sieboldiana* and *Hosta sieboldii*. *Hosta sieboldiana*, with large, grey-green leaves, with a blue-green edge, has been described as the 'archetypal hosta'; this and *Hosta sieboldiana* var. *elegans* are said to have good resistance to slugs. *H. sieboldii*, with white-margined or plain green leaves, makes good ground cover.

❧ Other garden plants Siebold collected include *Clematis florida* 'Sieboldii' (passion flower clematis), *Aucuba japonica* 'Variegata', and several hydrangeas, including *Hydrangea macrophylla* 'Maculata' and *Hydrangea paniculata* 'Grandiflora'.

TOP RIGHT *Hosta sieboldiana* was one of many hostas introduced to cultivation in Europe by Siebold. This clump-forming herbaceous perennial has white, lilac-tinged, bell-shaped flowers in summer.

RIGHT *Clematis florida* 'Sieboldii', introduced to England by Siebold in 1836, is notable for having stamens that have become petaloid. It is best kept in a sheltered spot.

John Lindley

DATE	1799–1865
ORIGIN	ENGLAND
MAJOR ACHIEVEMENT	*THE VEGETABLE KINGDOM*

Lindley, whose supreme talent as a botanist, horticulturist and organiser manifested itself in myriad administrative and academic roles, possessed formidable stamina and an appetite for incessant hard work. He published prolifically in the scientific and popular press, lectured constantly, giving the 1833 Royal Institution Christmas Lecture, and identified and named many plants, including Georgiana Molloy's specimens from Australia and William Lobb's giant redwood from North America. As consultant and campaigner, he saved the Royal Botanic Gardens, Kew, from threatened closure, and he campaigned against the glass tax that inhibited conservatory building.

Aeranthes grandiflora,
air flower

As a boy, Lindley collected wild plants as a hobby, and in his teens he became friends with William Jackson Hooker, another alumnus of Norwich grammar school. Lindley completed his first publication, *Observations on the Structure of Fruits and Seeds* (1819), a translation of Louis-Claude Richard's *Demonstrations botaniques, ou, L'analyse du fruit* at Hooker's home, working almost continuously on it for three days without going to bed. Hooker introduced Lindley to Robert Brown, botanist and librarian to Joseph Banks (see pp68–73), who in turn introduced him to Banks himself. Banks promised to send Lindley overseas as a naturalist, but later decided to employ him in his library and herbarium at Soho Square. By Banks's death, Lindley had completed important works on *Rosa*, aged 22, which he illustrated himself, describing 13 species for the first time, plus a monograph on *Digitalis* and a survey on *Pomoideae*. Lindley dedicated his monograph on roses to the botanist Charles Lyell, who presented him with £100 in return; Lindley used the money to purchase a microscope and a herbarium, which, by his death, contained around 58,000 specimens.

Lindley worked relentlessly, driven by financial circumstances – his father's nursery business had failed and Lindley had taken responsibility for his debts. His days began at dawn and continued into the evening, and he gave up to 19 different lectures a week. In order to keep up appearances, he was forced to borrow funds, too, living in debt to the banks and friends like Sir Joseph Paxton, with whom he shared a passion for orchids.

A PROLIFIC OUTPUT

Unable to refuse work, Lindley's output as an author and editor for the amateur gardener and botanist was significant, and he became one of the most important public figures in Victorian horticulture. He was an enthusiastic supporter of Antoine de Jussieu's 'natural system' rather than that of Linnaeus (see pp56–61), an early lecture proposing that botany should encompass 'the vegetable system in all its forms and bearings' to place plants in their natural families. These ideas were transcribed in *Synopsis of the British Flora* (1829) and *Introduction to the Natural System of Botany* (1830). Asa Gray, who reviewed the American edition, rejoiced, 'No book, since printed bibles were sold in Paris by Dr Faustus, ever excited so much surprise and wonder'. It was later revised and expanded, becoming *A Natural System of Botany* (1836). Among his arguments was the suggestion that names of families should end with '-acae', and orders with '-ales'. Lindley also published *The Vegetable Kingdom* in 1846.

Lindley was particularly passionate about orchids, writing *The Genera and Species of Orchidaceous Plants* (1830–40), *Sertum Orchidaceum* (1838), *Orchidacae Lindenianae* (1846) and *Folia Orchidacea* (1852–59). At a time when orchid imports from the tropics were in full flow and new, exotic plants, such as *Aeranthes grandiflora* (pictured on p86), needed names, Lindley established some 120 genera, becoming to many the 'Father of Orchidology'. Sarah Drake, a brilliant

botanical artist and one of the greatest orchid painters of all time, lived with the family from around 1830 to 1847, illustrating Lindley's work. Another collaborator was the geologist William Hutton; aware that transport and heating greenhouses depended on coal, Lindley's interest in fossil finds in coal seams encouraged him to compile *The Fossil Flora of Great Britain* (1831–37) with Hutton.

Although several of his works were accessible only to the wealthy, Lindley also wrote extensively for general readership, contributing 16,712 plant descriptions to John Loudon's *Encyclopaedia of Plants* (1829), alongside entries for the *Penny Cyclopaedia*, the *Dictionary of Science, Literature and Art* (1837) and Paxton's *Pocket Botanical Dictionary* (1840). Among his other works, *Ladies' Botany* (two volumes, 1837–38) sold well, but most successful of all was *Elements of Botany* (1841), a textbook that ran to numerous editions and translations. Lindley also edited *The Botanical Register* (1826–47), devoted to portraying new plants in colour, *The Journal of the Horticultural Society* (1846–55) and *The Gardeners' Chronicle*.

A LIFE OF ACHIEVEMENTS

In 1822 Lindley had overseen the construction of the Horticultural Society's garden at Chiswick, and by 1841 was its vice secretary. He and botanist George Bentham planned the first annual flower show for the organisation, and Lindley's collection of books became the basis of the RHS's Lindley Library.

As a consultant, Lindley advised the Board of Ordnance (on vegetable sources of carbon for gunpowder), the Admiralty (on the cultivation and reforestation of Ascension Island) and the Inland Revenue (on coffee and its adulterants). He also took part in a commission to investigate the Irish Potato Famine, having witnessed its effects first hand. In 1838 Lindley wrote an influential report on the Royal Botanic Gardens at Kew – at the time in a state of disrepair – recommending it become a national centre for botany, and thereby helping preserve it for future generations.

Rosa fraxinifolia (Newfoundland rose), collected by Sir Joseph Banks and painted by Lindley, who was not only a scientist but also a talented artist.

Lindley was elected to the Linnean and the Geological societies in 1820, to the Imperial Academy of Natural History in Bonn in 1821, and to the Royal Society in 1828 (and awarded its royal medal in 1857). His work was honoured with the naming of the genus *Lindleya* (1824). In 1829 Lindley became the first Professor of Botany at the University of London, a post he held until 1860. By retirement he had authored over 200 publications.

In his rare moments of leisure, Lindley gardened and enjoyed archery and rifle shooting, having a 100-yard range in his garden. Although reputed to be a good shot, he once missed and hit his servant in the thigh. (He later returned the compliment, accidentally striking Lindley on the head with an axe handle.)

Lindley died of apoplexy on 1 November 1865, shattered by ill health and exhaustion – just outlasting his friends Sir Joseph Paxton (8 June 1865) and Sir William Hooker (12 August 1865). He was buried in Acton cemetery five days later.

John Lindley:
INSPIRATION FOR GARDENERS

❧ One genus established by Lindley was *Chaenomeles* (Japanese quince), excellent free-standing or wall-trained deciduous plants, flowering in early spring before leaves appear. *Chaenomeles speciosa* has red flowers, followed by small green, fragrant fruits; *C. speciosa* 'Nivalis' has pure white flowers with pink tints; *C. speciosa* 'Moerloosei' has white flowers tinged with deep pink.

❧ Lindley also named the genus *Eriobotrya* and the species, *Eriobotria japonica* (loquat), a large shrub or medium-sized evergreen tree with large glossy leaves and clusters of small fragrant flowers, followed by small edible fruit that rarely ripen in Britain, but do in

BELOW Lindley established the genus *Chaenomeles*, which includes *Chaenomeles speciosa* (Japanese quince), a colourful shrub that is easy to grow.

Mediterranean climates. Since it is not reliably hardy, grow it in a warm, sheltered position or by a hot sunny wall. This plant was introduced into cultivation in Britain through Kew in 1787, at the time of Sir Joseph Banks.

✤ Lindley named two *Phalaenopsis* species in the *Gardeners' Chronicle* of 1848 – *Phalaenopsis rosea* (pink butterfly plant), introduced by collector Thomas Lobb from Manila through the Veitch Nurseries, and *Phalaenopsis grandiflorum* (now *P. amabilis*), which Lobb sent from Java. Both men would have been delighted at how selections of this genus have become widely grown as houseplants. They need good light in winter; a bright windowsill is ideal. Remove them from the windowsill overnight before closing the curtains, and provide shade from scorching sunshine in summer. Temperatures at night should range from 16°C to 19°C (60–66°F) and day temperatures should be between 19°C and 30°C (66–86°F), avoiding fluctuations and draughts. When flowers fade, cut off the spike to the joint below on the flowering stem and a new shoot will develop. If this does not happen, reduce temperatures by 5°C (41°F) to initiate flowering. Water regularly through the growing season, and reduce watering in winter. Do not wet the leaves, but mist them occasionally with tepid water in summer. The roots should not remain constantly wet, but neither should they dry out completely. Feed the plant when you water it apart from every fourth time, when the pure water will prevent the build-up of excess salts in the compost. Feed occasionally in winter and repot in spring.

ABOVE RIGHT Lindley named a new genus and species when he described *Eriobotrya japonica* (loquat), which will fruit after long hot summers.

RIGHT *Phalaenopsis amabilis* (East Indian butterfly plant), named by Lindley, has become a 'must have' orchid for the home.

Anna Atkins

DATE	1799–1871
ORIGIN	ENGLAND
MAJOR ACHIEVEMENT	*PHOTOGRAPHS OF BRITISH ALGAE*

The botanist Anna Atkins became famous for perfecting the art
of the cyanotype, an early photographic printing process, and producing
the first book in the world to be illustrated by photographs. She began
by producing images of seaweed gathered around the coasts of Britain,
or sent to her by friends, moved on to ferns, and then became
accomplished at portraying flowering plants. Her artistic eye and the
natural elegance of her subjects produced works that combined the best
in the world of fine art, natural science and early photography.

Phegopteris connectilis,
beech fern

After her mother died shortly after childbirth, Anna and her father, John Children, became very close and shared the same interests. It was from him that Anna gained a love of science; unusually for the time, he believed that gender was no barrier to success and encouraged her interest in botany. Through his position as a librarian at the British Museum and Secretary of the Royal Society, he encouraged her participation in what were traditionally male-dominated circles. His contacts in the scientific world were useful, too. He had a large, well-equipped scientific laboratory at his home in Tonbridge, and there is an account of a meeting there, in 1813, of 38 leading English chemists (including Humphry Davy), who had gathered to investigate the properties of a large battery that Children had made. Anna was already a member of the Botanical Society of London when her father chaired the February 1839 meeting of the Royal Society at which William Henry Fox Talbot first revealed the details of photography. A few months later, Children wrote to Talbot, 'my daughter and I shall set to work in good earnest 'till we completely succeed in practicing your invaluable process'.

In 1842 Sir John Herschel sent Children a copy of his paper describing his refinement of Talbot's process, an invention he called the cyanotype. Making a cyanotype was relatively simple. A piece of plain paper was coated with light-sensitive iron salts known as 'Prussian blue', in a darkened room; an object was then laid on the paper, exposed to light for around 20 minutes, depending on the strength of the sun, during which time the exposed paper reacted with the light, turning brilliant blue, while the covered parts remained white. The image was then fixed by washing the paper in clean water, to which a little sulphate of soda had been added, creating a 'blueprint'. It was inexpensive, relatively easy to master and the blue image tone was perfect for plants. Children's laboratory and its giant battery probably also provided a source of the ammonium ferric citrate and potassium ferricyanide needed for the process.

THE POSSIBILITIES OF PHOTOGRAPHY

Anna was a knowledgeable botanist but a skilled illustrator, too. After contributing more than 200 detailed drawings to her father's 1823 translation of Jean Lamarck's *Genera of Shells*, however, she realised that photography could provide a time-saving method of producing illustrations. Botanists had already seen the potential of photography for recording complex scientific specimens, but it was Anna who thought of them in terms of publication.

Anna was so disappointed by the lack of illustrations in William Harvey's 1841 *Manual of the British Algae* that she decided to produce her own, using specimens from her own collections. *Photographs of British Algae: Cyanotype Impressions*, including 398 plates and 14 pages of text, became the first book in the world to be illustrated with photographs. Each was a perfect example of its kind, artistically laid out, showing the contrasting forms among seaweeds such as *Cystoseira fibrosa* (first named by Linnaeus), with its filamentous stems and tiny flotation bladders. For this

Cassia
(America)

Pteris aquilina

Spirea aruncus
(Tyrol)

Cystoseira granulata

specimen, the stems were laid out in a graceful arc to illustrate their flexibility. The presentation of *Sargassum bacciferum* was similarly elegant. The book was published in three volumes between October 1843 and September 1853 and distributed to scientific friends and institutions in paper wrappers and unbound – recipients were expected to undertake the binding themselves. Talbot expressed his gratitude for her 'photographing the entire series of British seaweeds, and most kindly and liberally distributing the copies to persons interested in Botany and photography'. Anna likewise wrote in the text accompanying the book: 'The difficulty of making accurate drawings of objects so minute as many of the Algae and Confervae has induced me to avail myself of Sir John Herschel's beautiful process of Cyanotype, to obtain impressions of the plants themselves, which I have much pleasure in offering to my botanical friends.' Based on the number of surviving copies, she produced 13 editions, describing them as 'impressions of the plants themselves'. She used good-quality paper, so the images are still in excellent condition today, and she became adept at judging how long the paper needed to be in sunlight to produce images that were as clear as possible. Every one of the images in her books was produced by her hand. At the time, cyanotypes were more permanent than early photographs, so her work remained in good condition for many years.

Atkins's cyanotypes (clockwise from top left): *Carex* (sedge), *Pteridium aquilinum* (bracken), *Cystoseira granulata* and *Aruncus dioicus* (goat's beard). Her images are still in great condition today as she used high-quality paper and was so careful when judging how long to expose them to sunlight.

FURTHER WORKS

Anna later collaborated with fellow female botanist and childhood friend Anne Dixon (second cousin of the novelist Jane Austen) to produce two more books using cyanotypes. The first was *Cyanotypes of British and Foreign Ferns* (1853), which included specimens such as *Woodwardia virginica*, *Pteris rotundifolia*, *Adiantum serrulatum*, *Polypodium phegopteris* (now known as *Phegopteris connectilis*, pictured on p92), *Polypodium dryopteris* and *Aspidium denticulatum*.

Cyanotypes of British and Foreign Flowering Plants and Ferns followed in 1854 and included ethereal images of the flowers of *Papaver rhoeas*, *Papaver orientale*, *Leucojum aestivum* and *Taraxacum officinale*. Anna also experimented with other objects, such as feathers and pieces of lace. The text was presented as photographic copies of her own handwriting, written on paper with pen and ink. The paper was then coated in oil to make it transparent, laid on the treated paper, then moved into the light. When the paper turned blue, the top sheet was removed.

The works of Anna Atkins were hugely significant events in the world of botany and photography, and originals are now very rare; the last on sale raised 450,000 euros at auction. After a life in which she created at least 10,000 images by hand, Anna died at Halstead Place in Sevenoaks, Kent, on 9 June 1871 and was buried in the local churchyard there.

Anna Atkins:
INSPIRATION FOR GARDENERS

❧ One of the ferns imaged by Anna Atkins was *Woodwardia virginica* (American chain fern) from part-shaded acidic swamps and wetlands in coastal Canada and the eastern seaboard of the United States. The young fronds of this deciduous fern are reddish brown, reaching 90cm (35in) long in average conditions, or up to 3m (10ft) in boggy ground. This creeping fern can be invasive where conditions replicate its native habitat in wet mud or even underwater. It can also be planted in full sun, providing there is sufficient water. Young plants grow quite well in moist

BELOW In 1854, Atkins turned to wild flowers. Her cyanotypes included reliable border plant *Leucojum aestivum* (summer snowflake).

garden soil but generally die during winter. The rhizomes are tender and deep water protects them from severe frost, so a thick mulch of well-rotted organic matter should improve the chances of success.

❧ *Leucojum aestivum* (summer snowflake) carries beautiful heads of white, green-tipped flowers above elegant leaves in spring, just after the daffodils finish flowering. It likes moist soil and is even happy by the margins of a pond. *Leucojum aestivum* 'Gravetye Giant' was introduced by the great gardener William Robinson, probably from around Limerick, in Ireland, where a particularly large-flowered form is known to be found.

✤ Atkins took a cyanotype of *Papaver rhoeas*, the common poppy associated with the fields of Flanders. In 1880, the Reverend William Wilks discovered a single flower with a white margin to the petal, picked the seed capsule, then spent 20 years breeding soft-coloured white, pink, yellow and orange selections, some with dark edges (picotees); he named them 'Shirley' poppies after his Surrey parish. Other forms, collectively known as 'Mother of Pearl', were selected after the Second World War by artist and gardener Sir Cedric Morris, who

ABOVE Among Atkins's cyanotypes was one of *Papaver rhoeas* (common poppy). Hardy annuals like poppies are ideal as gap fillers in gardens.

searched Suffolk for soft tints like raspberry-pink, blue-grey and soft lilac. Today several similar single strains are available, such as 'Mother of Pearl', 'Cedric Morris' and 'Fairy Wings'; selections of semi-doubles and doubles include 'Angels' Choir' and 'Shirley Double Mixed'.

Georgiana Molloy

DATE	1805–1843
ORIGIN	ENGLAND
MAJOR ACHIEVEMENT	BOTANICAL COLLECTOR

Georgiana Molloy is remembered as one of the first botanists in the Swan River Colony in Western Australia. As a retreat from the drudgery of her arduous daily life as an early settler, she spent her time gardening and collecting botanical specimens in an area scientists now consider to be one of 25 biodiversity hotspots in the world. What began as a hobby ended up as a work of great scientific value, and her collections were regarded as the finest of their day to arrive in Britain.

Kennedia carinata

Georgiana's early childhood was idyllic. She lived in Crosby Lodge near Carlisle, England, in a country house with a large kitchen garden, surrounded by open countryside, where she became passionate about gardening and pressing wild flowers. Then, aged 14, disaster struck. Her father died after falling from a horse, leaving a widow and five children with considerable debts. When the family relocated to 'horrid Rugby' in Warwickshire, Georgiana broke with the strict traditions of the time, leaving home without the prospect of marriage or employment, and going to stay with friends at Keppoch House in Dunbartonshire. She remained there for 18 months, seeking solace among plants.

While in Scotland, a long-standing friend, the career soldier Captain John Molloy, exchanged his post with another officer, went onto half-pay, was listed as unattached from his regiment, and decided to join the *Warrior*, sailing for southwest Australia. The Swan River Colony (now the city of Perth) was its destination – the first non-convict colony in Australia. Molloy proposed to Georgiana in 1829 and, several months later, they set off on the demanding five-month voyage. Packed in her luggage were plants, seeds and a collection of pressed flowers she had accumulated since 1821.

A PASSION FOR COLLECTING

Life as a pioneer was punishing. At the time, the colony was a collection of tents, crowded huts and livestock, so John and Georgiana (who struggled with the heat, fleas, flies and dysentery) moved 320km (200 miles) south along the coast to Augusta. This was, however, the beginning of further hardship and tragedy. Georgiana's first daughter, born in a tent on the beach soon after landing in Augusta, died a few days later; her only son drowned in a well, aged one-and-a-half, and the colony almost starved to death when supplies failed to arrive. In between, Georgiana endured hours of loneliness while her husband worked away from home.

Yet again Georgiana sought comfort among plants, creating a garden with English flowers and seeds from the Cape, but she soon found greater joy in the native wild flowers that bloomed profusely in spring (such as the southwest Australian *Kennedia carinata*, pictured on p98). At first, she pressed and mounted them, sending them to family and friends in her letters. However, her life changed dramatically in 1838 when Lady Stirling, wife of the Governor of Western Australia, suggested to her cousin, Captain James Mangles, that he ask her to send plant material to him – he was particularly interested in seeds.

Georgiana then spent the last six years of her life pressing plants, describing them and collecting seeds; sending seeds and plants to London became her passion. She even persuaded others to collect material for her: her husband, daughters, farm workers – even soldiers walking between the towns of Vasse and Augusta – brought her specimens. But it was Georgiana who dried and labelled them all. Her pressed specimens were mounted with meticulous care, and her

plant descriptions were known for their great detail – habitat, flowering time, soil and amount of moisture required. Georgiana waited patiently for seeds to ripen, and her samples were renowned for their freshness and careful packing. Each plant was mounted in a book with a corresponding number on the seed packet, and each package contained a detailed explanatory letter. As she wrote on one occasion: 'I have minutely examined every seed and know they are sound and fresh as they have all been gathered in the past five weeks during December and January.' A parcel sent in 1841 contained 100 different species.

In England, James Mangles dispersed the pressed specimens and seeds he received to 15 of the greatest scientists and gardeners of the day – including Joseph Paxton at Chatsworth House, and George Loddiges, whose nursery was famous for its hot-house orchids. In the words of Paxton, Georgiana's collection was 'collectively, the best and contains more good things than I have before received from that interesting part of the world'; he described her 'important collection of seeds' as 'far superior to any we have received at Chatsworth'.

RECOGNITION AT LAST

The botanist John Lindley (see pp86–91), who was fascinated by Australian flora (particularly its orchids), was very impressed by Georgiana's work. In his appendix to the first 23 volumes of *Edwards's Botanical Register*, which included a sketch of the vegetation of the Swan River colony, he wrote glowingly of the lady 'enthusiastically attached to the Botany of this remote region'. To Mangles he wrote: '... Mrs. Molloy is really the most charming personage in all South Australia and you the most fortunate man to have such a correspondent. That many of her plants are beautiful you can see for yourself & I am delighted to add that many of the best are quite new. I have marked many with a X.' Yet despite declaring that Georgiana's 'zeal in the pursuit of Botany has made us acquainted with many of the plants of that little known part of the world', her name was omitted from Lindley's list of acknowledgements.

On 6 June 1830, Georgiana collected what is presumed to be *Sollya heterophylla* (bluebell creeper), and placed it in her baby's coffin.

Georgiana and Captain Mangles's shared passion for plants, however, led to a mutually rewarding correspondence. In a letter of 1 February 1840, she wrote: 'Our Acquaintance is both singular and tantalizing, and somewhat melancholy to me, my dear Sir, to reflect on. We shall never meet in this life. We may mutually smooth and cheer the rugged path of the World's Existence, even in its brightest condition, by strewing flowers in our Way, but we never can converse with each other, and I am sincere when I say, I never met with anyone who so perfectly called forth and could sympathize with me in my prevailing passion for Flowers.'

Georgiana died in childbirth in 1843 aged just 37, having compiled around 1,000 collections. Late in her life she wrote to Mangles, 'I have sent you everything worth sending'. This talented botanist-gardener is now commemorated by the beautiful, pink-flowered shrub *Boronia molloyae* (tall boronia).

Georgiana Molloy:
INSPIRATION FOR GARDENERS

✤ A 'British' garden could not survive the long, dry summers and poor soil of southwest Australia, so Georgiana developed her own ideas and style, combining plants – particularly natives – with plants from other parts of the world. The results were the talk of the colony. Assess the conditions, then adapt to achieve success.

BELOW *Boronia molloyae*, named for Georgiana, is also called 'tall boronia'. Molloy's husband once saw it growing 'as high as his head, riding on horseback', according to Sir Joseph Hooker.

✤ Georgiana found ways of adapting, despite the limitations of what was available. When fungus appeared in some seeds, she rinsed them in salt water, using its antifungal properties; as lime was unavailable, she used wood ash, with an alkaline pH instead. Be flexible, resourceful and unafraid to experiment; sometimes this is the only way to create a garden.

✤ At the start, Georgiana didn't really know how to collect, mount or pack herbarium specimens. Lacking a plant press, she dried plant material between sheets of paper and weighted it down. Until Mangles sent her absorbent paper, everything she used would have been a great sacrifice. Bernice Barry, an authority on Georgiana, suggests that she probably used blotting paper, as writing letters and keeping a journal would have been a daily essential for her husband. Georgiana owned few books so it is likely she used rocks as weights. If you want to try pressing flowers, do so at their peak, once the dew has evaporated from the flowers. If you cannot press them immediately, store specimens in a ziplock bag, in the fridge.

✤ Georgiana noted where she had found plants, even marking the plants themselves, then returned later for the seeds. She made bags to keep seeds separate and discarded any that had been entered by even the tiniest insects. She also experimented with ways to prevent infestation – sprinkling seeds with pepper or wrapping them in 'tanner's bark', and later sending living plants in a 'Wardian case'. When collecting seeds at home, cut the capsules just before they ripen and pop them into a paper bag. Put this in a warm, dry place and the shed seed will be easy to collect later.

TOP *Beaufortia squarrosa* (sand bottlebrush) grew where Molloy lived in Vasse, Western Australia. In Britain, this shrub needs a conservatory.

ABOVE Molloy was a fervent collector; she urgently searched for *Nuytsia floribunda* (Australian fire tree) seeds before she died.

Asa Gray

DATE 1810–1888	
ORIGIN UNITED STATES OF AMERICA	
MAJOR ACHIEVEMENT *GRAY'S MANUAL*	

Considered to be the most important American botanist of the
nineteenth century for what became known as *Gray's Manual*, Asa Gray
was a close friend and correspondent of botanist Sir Joseph Hooker,
naturalist Charles Darwin and many of the other great scientists of the
time. A committed Christian yet also a passionate supporter of Darwin's
ideas, his *Darwiniana* – a collection of scientific and philosophical papers
– was instrumental in introducing the theory of evolution in North
America in a non-controversial and academically reasoned manner.

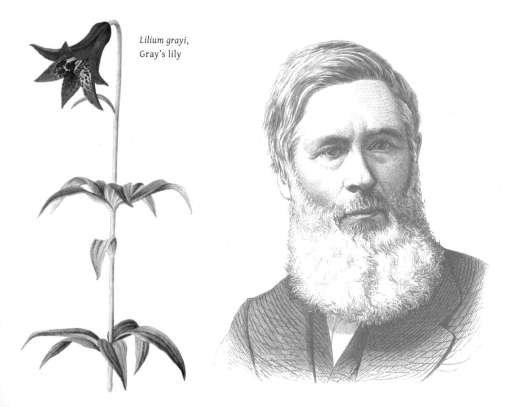

Lilium grayi,
Gray's lily

One winter, partway through a course at the Fairfield Medical College in New York, Asa Gray read an article in Brewster's *Edinburgh Encyclopaedia*, which fired his interest in botany. The following spring he eagerly watched for the first flower to appear and, with the aid of Amos Eaton's *Manual of Botany for Northern and Middle States*, found it to be *Claytonia caroliniana* (Carolina springbeauty). Gray had discovered his passion. Correspondence with the eminent botanist Dr Lewis C. Beck about plants that he collected led to a meeting with Dr John Torrey, Professor of Chemistry and Botany at the New York College of Physicians and Surgeons, with whom Gray had been corresponding since 1 January 1831. First, Gray became Torrey's assistant and collector. Then in 1836, through Torrey's kindness, Gray became Curator of the New York Lyceum of Natural History. His first botanical papers, entitled 'A Notice of some new, rare, or otherwise interesting Plants, from the Northern and Western portions of the State of New York and A Monograph Of the North American Species of Rhynchospora' were read before the Lyceum in December 1834.

Gray was later appointed botanist on an expedition to explore the South Pacific. However, excessive delays led to him resigning his commission to collaborate with Torrey in compiling *A Flora of North America*. On a visit to Europe, for future volumes, Gray consulted herbaria holding large collections of American plants in seven countries. He also met many eminent botanists and formed lifelong friendships, including one with Sir Joseph Hooker, before returning to America well equipped to continue the work.

In 1842, he became Fisher Professor of Natural History in Harvard University, a position he occupied until his death. He spent his time researching and developing the botanic garden there. When he arrived there was one greenhouse, no herbarium and there had been no curator for eight years. Gray donated his library and specimens to what is now known as the Gray Herbarium and revived the garden.

GRAY'S MASTER WORK

Gray's classic work, first entitled *A Manual of the Botany of the Northern United States from New England to Wisconsin and South to Ohio and Pennsylvania Inclusive*, ran to eight editions and gradually increased its geographical coverage. The single volume was too large to be the portable excursion flora Gray had hoped for (though enthusiasts remained undeterred). In 1845 Gray wrote to Torrey that he was writing an 'imperfect and hasty' botanical manual, to be used alongside his *Elements of Botany* (1836) and *The Botanical Text-book* (1842). Two years later he focused almost all his efforts on the project, putting himself heavily in debt in terms of both time and money. When it was finally published, he wrote in frustrated tones: 'What can you expect from a man who takes up a job in February 1847, to finish in May or June certain; but who, though he works like a dog and throws everything else, does not get it done until February comes round again.' It took Gray just a year to write a 783-page manual, though he intended it to be half

that size. The impact on the botanical world was immediate and lasting, awakening or intensifying the passion of botanists and lovers of plants around the globe.

Sequoia sempervirens (coastal redwood). The history of the sequoia was the subject of a lecture given in Iowa by Gray in 1872.

Gray had many correspondents, the most enduring of these being Charles Darwin, and letters between the pair mostly discussed design in nature and religious belief. Darwin kept Gray well informed about the publication of *On the Origin of Species*; Gray in turn became one of America's most active promoters of evolution while retaining his belief in a Creator. Gray, a devout Christian, believed that natural selection created new species, but that it was not the only cause of variation; he believed that an inherent power had also been imparted in the beginning by God. Gray's many supportive essays in scientific journals, were published as *Darwiniana: Essays and Reviews Pertaining to Darwinism* (1876). As Darwin noted to A. R. Wallace, 'Asa Gray fights like a hero in defence'.

THE FATHER OF AMERICAN BOTANY

Remarkably, the man who described so many new American species did not travel west of the Mississippi River until the age of 62, working instead from herbarium specimens: 'Although I have cultivated the field of North American botany ... for more than 40 years ... yet so far as our own wide country is concerned, I have been to a great extent a closet botanist.'

In 1873 he retired to work on *A Flora of North America*, the last volume having been published 30 years before. He travelled extensively throughout North America, from Mexico to Montreal, including a trip to the Rocky Mountains with Sir Joseph Hooker (see pp116–121), as guests of the United States Geological and Geographical Survey. A photo of 25 July 1877 shows the party camped at La Veta Pass, Colorado, with Hooker on a chair, and Gray kneeling on the ground beside him, holding a full herbarium press, both deep in conversation.

On Gray's 75th birthday, the botanists of North America, led by the editors of the *Botanical Gazette*, presented him with a silver vase, inscribed '1810, November eighteenth, 1885. Asa Gray, in token of the universal esteem of American Botanists', and embossed with flowers relevant to the occasion, including *Grayia polygaloides*, *Lilium grayi* (pictured on p104) and *Notholena grayi*. A silver salver, bearing the cards of the contributors, was also given, marked with the inscription, 'Bearing the greetings of one hundred and eighty botanists of North America to Asa Gray on his 75th birthday, Nov. 18, 1885'.

Upon his death in 1888, having done more than anyone to unify the taxonomic knowledge of North America, Gray was lauded as 'one of the most remarkable men of our country, and as a scientist, the best and most esteemed abroad of any American of our day'.

Sequoiadendron giganteum (big tree), one of the trees seen by Gray during his travels for the Geological Survey.

Asa Gray:
INSPIRATION FOR GARDENERS

✤ Asa Gray and Sir Joseph Hooker visited the Californian redwood forests, where they were shown two genera described as 'the vegetable wonders of the world' – *Sequoia sempervirens* (coastal redwood) and *Sequoiadendron giganteum* (big tree). The latter was introduced by plant collector William Lobb at the height of the gold rush. In 1852, Lobb set out for Calaveras Grove, California, where he found around 80 trees ranging from 71 to 76m (233–250ft) tall, and 7.6m to 9m (25–30ft) in diameter. He packed his saddlebags with seeds and two seedling plants and, gambling on

incurring the wrath of his employers at the Veitch Nurseries, yet convinced of the treasure he'd found, cut short his collecting trip and returned home to England with his bounty. He arrived back on 15 December 1853. On Christmas Eve the 'big tree' was given the name *Wellingtonia gigantea* by John Lindley, after the Duke of Wellington who had died the previous year. This angered American botanists (who had proposed the name *Washingtonia gigantea*); they were outraged that the world's largest tree had been named for an English war hero, by a botanist who had

BELOW Gray felt that care should be taken over plants' common names, writing that *Clematis virginiana* (woodbine) should be called 'virgin's bower', as it was not fitting for another plant.

BELOW On their travels, Gray and Hooker saw *Sequoiadendron giganteum* (big tree), which was later introduced commercially to Britain from seed and widely planted as status symbols on country estates.

ABOVE *Carex grayi* (mace sedge), named for Gray, is an attractive garden plant for moist soil and works well by the edge of a pond.

not even seen the tree. Eventually, it was given the generic name *Sequoiadendron*, for Sequoia (George Gist), the son of a German-American merchant and the Native American girl who invented the Cherokee alphabet. The first plantings of these trees are now emerging above the surrounding trees in Britain, just as John Lindley predicted. They are well worth a visit at Killerton Garden, Devon, Sheffield Park and Garden or Penrhyn Castle, Wales. In California, conservationists – notably John Muir – have also ensured these great trees' survival. If a *Sequoiadendron giganteum* is too large for your garden, try *Metasequoia glyptostroboides* (dawn redwood), a deciduous conifer for moist soil. Its leaves are fresh green in spring and bronze in autumn, and it has attractive bark and a pleasing winter silhouette. Avoid locations with late frosts.

RIGHT/CENTRE RIGHT *Diphylleia grayi*, named for Gray and in the Berberidaceae family, is notable for having white petals that become translucent in the rain. Its pretty flowers are followed by dark-purple fruits.

William Colenso

DATE	1811–1899
ORIGIN	ENGLAND
MAJOR ACHIEVEMENT	MISSIONARY, BOTANIST AND EXPLORER

William Colenso, a Cornish missionary to New Zealand, combined preaching with botany. During his lifetime he walked thousands of miles collecting plants as he visited his parishioners. Colenso met and corresponded with many great botanists of the time, including Joseph Hooker, often accompanying him out into the 'bush' to search for plants on his visit to New Zealand. A man of robust physique, an active and retentive mind and boundless enthusiasm, Colenso visited the spots where Banks and Solander had collected plants nearly 70 years before and spoke with Māori who remembered Captain Cook.

Metrosideros excelsa,
New Zealand iron tree

William Colenso showed an early interest in natural history, reading his first paper to the Penzance Natural History and Antiquarian Society when he was 18. First apprenticed to a printer, he then moved to London and later became a missionary/printer at the Paihia missionary station in the Bay of Islands, in New Zealand's North Island.

Once in New Zealand, Colenso benefited from the encounter with two eminent botanists – Charles Darwin in 1835 and Allan Cunningham, the government botanist of New South Wales, in 1838. Cunningham remained in New Zealand for several months, and during this time he guided Colenso's interest in natural science towards botany, teaching him how to collect plants. In a letter to Colenso, Cunningham urged him 'Not to lose sight of the vegetation of the land you live in, and do not scatter to the winds that little you gather'd regarding the peculiarity of those vegetables, when I was with you'.

In 1839 Colenso travelled extensively through the North Island, gathering knowledge of its natural history and discovering two small Australian plants in New Zealand, a sundew (*Drosera pygmaea*) and a club moss (*Lycopodium drummondii*); neither was rediscovered for over 50 years. He also saw *Lobelia physaloides* in the northern part of the North Island, which Sir Joseph Hooker later classified as *Colensoa physaloides*, a blue-flowered, single-species genus found nowhere else in the world.

EXPLORING THE NEW WORLD

Prior to the British Antarctic Expedition spending three months in the Bay of Islands in 1841, Sir William Hooker advised his son Joseph, who was onboard (see pp116–121) to seek out Colenso upon his arrival. Hooker admired the collection of minerals, shells and insects belonging to this 'brisk and active man' and the New Zealand flora growing in his garden; Colenso often acted as Hooker's guide as they explored the native bush together. Hooker wrote to his father: 'Colenso has been extremely kind to me ... He is a very good fellow in every respect, and has shown me the greatest attention ... Of this class of men Mr Colenso is among the most superior.' Afterwards, the two men continued to correspond for over 50 years, further encouraging Colenso's interest in the plants of his adopted home. Colenso sent large shipments of specimens to the Hookers at Kew Gardens for 14 years, and over 800 herbarium sheets at Kew are attributed to him. He also retained a collection in New Zealand.

Colenso collected wherever he went. In the summer of 1841 he travelled to the east coast, and when he reached Poverty Bay, he found *Coriaria kingiana*, a low-growing bush with undulating leaves, and *Metrosideros excelsa*, the red-flowered New Zealand iron tree (pictured on p110), or pohutukawa tree, usually associated with the seashore, along the edge of the freshwater Lake Waikaremoana, 80km (50 miles) west–southwest of Gisborne. He also found several 'filmy' ferns deep in the forest – some new to science – including *Hymenophyllum pulcherrimum*, or tufted filmy fern, and *Dicksonia lanata*, the stumpy tree fern.

In 1843, a Māori, paid by Colenso, climbed to the bush line on Mount Hikurangi, returning with a collection of alpines, including a large yellow buttercup (*Ranunculus insignis*), a mountain daisy with white woolly leaves (*Celmisia incana*) and North Island edelweiss (*Leucogenes leontopodium*). A year later he saw New Zealand alpines for himself at the summit of the Ruahine Range, which runs parallel with the east coast of the North Island between East Cape and Wellington. He wrote, 'When we emerged from the forest and the tangled shrubbery at its outskirts on to the open dell-like land just before we gained the summit, the lovely appearance of so many and varied beautiful and novel wild plants and flowers richly repaid me the toil of the journey and ascent, for never did I behold at one time in New Zealand such a profusion of Flora's stores. In one word, I was overwhelmed with astonishment, and stood looking with all my eyes, greedily devouring and drinking in the enchanting scene before me ... Here were plants of the well-known genera of the bluebells and buttercups ... daisies, eyebrights and speedwells of one's native land, closely intermixed with the gentians of the European Alps, and the rarer southern and little-known novelties – *Drapetes, Ourisia, Cyathodes, Abrotanella,* and *Raoulia.*'

Colensoa physaloides, originally named for Colenso, is now classified as *Lobelia physaloides*. A New Zealand native, it was once widely distributed in coastal and lowland forests, but it is now only common on some offshore islands.

Colenso gathered as much as he could, but was unprepared for such bounty, writing, 'But how was I to carry off specimens of these precious prizes, and had I time to gather them? These mental pictures completely staggered me, for I realized my position well. We had left our encampment that morning, taking nothing with us, so we were all empty-handed.... I had no time to lose, I first pulled off my jacket, a small travelling-coat, and made a bag of that, and then, driven by necessity, I added thereto my shirt, and by tying the neck, etc., got an excellent bag; whilst some specimens I also stowed in the crown of my hat.'

When later given an extensive parish, Colenso toured it twice a year, on foot. His most notable discovery on these tours was the shrub *Senecio greyi* (now *Brachyglottis greyi*) on the east coast north of Cape Palliser, long a favourite in gardens. He also published over 50 botanical papers, most in *Transactions and Proceedings of the New Zealand Institute*, of which he was a founder member, and when Hooker produced his *Flora Novae-Zelandiae* in instalments between 1853 and 1855, three botanists, including Colenso, had contributed richly to his knowledge in the field, through herbarium specimens and correspondence.

Colenso was elected a Fellow of the Linnean Society in 1865 and the Royal Society in 1886. To the latter, he wrote of the 'high honour' that had been conferred upon him and of his hope that it 'may serve to stimulate me yet more in pursuing in the paths of science'. He died in Napier, on the east coast of the North Island, on 10 February 1899.

William Colenso:
INSPIRATION FOR GARDENERS

❖ Twenty-six New Zealand plants have been named in Colenso's honour, among them *Hebe colensoi*, with greyish-green leaves, and *Olearia colensoi*, a daisy bush with thick serrated leaves. Both genera are a source of excellent garden plants. Hebes have been bred extensively and are valued for being evergreen, compact in growth and often long-flowering. Among those with the Awards of Garden Merit from the RHS are low-growing *Hebe pinguifolia* 'Pagei', with grey foliage and short spikes of flowers in spring and early summer; *Hebe* 'Sapphire', which has dark green foliage and lavender flowers in summer; and *Hebe* 'Great Orme', found as a chance seedling in north Wales, being compact and rounded, with bright pink flowers fading to white.

❖ Colenso also sent pressed specimens of the genus *Olearia* to Kew. The daisy bush makes an excellent wind- and salt-resistant plant for mild and maritime climates. *Olearia macrodonta*, known as New Zealand holly, has spiny grey-green leaves with a white felted underside and small clusters of white flowers, while *Olearia cheesemanii* produces billowing clouds of white daisy flowers in spring that almost cover the plant. One of the most beautiful of the genus, *Olearia* 'Henry Travers', found in the wild on the Chatham Islands, is medium-sized, with grey-green leaves that are silvery below, and daisy-like flowers with a purple centre and lilac petals in midsummer.

BELOW Colenso collected several species of hebe, which have become a familiar garden plant. *Hebe* 'Sapphire' is a small, rounded evergreen with spikes of flowers in summer.

ABOVE In 1849, Colenso sent a pressed specimen of
Olearia macrodonta (New Zealand holly) to the Royal
Botanic Gardens, Kew, where it still remains.

RIGHT Among the specimens Colenso pressed was
Astelia nervosa (bush flax). 'Westland' (pictured) is
one of the best cultivars for borders or containers.

❖ Two herbarium sheets of *Astelia nervosa*
(bush flax) at the Royal Botanic Gardens, Kew,
were sent by Colenso. This is a tussock plant
with sword-shaped silvery leaves up to 2m
(6¹/₂ft) long and 4cm (1¹/₂in) wide, forming
large colonies at altitude on damp grasslands, fell
and herb fields; its flowers are small but sweetly
scented. The selection 'Westland' has silvery
bronze leaves that often adopt red tints in cold
weather. It needs a sheltered position in moist,
humus-rich soil, and may need frost protection
in winter in cooler climates.

Joseph Dalton Hooker

DATE 1817–1911	
ORIGIN ENGLAND	
MAJOR ACHIEVEMENT BOTANIST AND EXPLORER	

Sir Joseph Dalton Hooker was a man of many talents – botanist, plant hunter, phylogeographer, artist and cartographer. Gardeners remember him for the multitude of rhododendrons from Sikkim that he introduced to the grand estates of his homeland, transforming the style of British gardens. He was a close friend and supporter to many, a collaborator to Charles Darwin, and, in his capacity as Director of the Royal Botanic Gardens at Kew, responsible for establishing botanic gardens throughout the British Empire. No wonder, then, that he is regarded as one of the greatest botanists, naturalists and explorers of the nineteenth century.

Rhododendron ciliatum

Joseph Dalton Hooker was born in Halesworth, Suffolk, on 30 June 1817. His father, Sir William Hooker – who would later become the first Director of the Royal Botanic Gardens at Kew – was a botanist and professor at Glasgow University, so the young Hooker spent his formative years in that town. Even as a child he had a keen interest in plants. When aged around 5 and found 'grubbing in a wall in the dirty suburbs of the city of Glasgow', as he later put it, he simply explained that he had found the moss *Bryum argenteum*.

As a boy, Hooker was enraptured by the idea of exploration, avidly reading Mungo Park's *Travels in the Interior Districts of Africa* and an account of Captain Cook's voyages. Then, at the age of 15 he began to study medicine at Glasgow University, during which time he became friends with Charles Darwin, whom he met by chance in Trafalgar Square. Hooker graduated in 1839. In that same year, he learned that Captain James Clark Ross was about to lead an expedition to the Antarctic, and Hooker was determined to be on board.

On 28 September 1839, HMS *Erebus* and HMS *Terror* left England, with the 22-year-old Hooker as Assistant Surgeon (though he preferred the title 'Botanist to the expedition'). Their journey was not without incident. On the night of 13 March 1842, Ross recorded, the ships encountered a tempestuous gale and a giant iceberg. Ross wrote that the two ships became 'entangled by their rigging, dashing against each other with fearful violence … Sometimes the "Terror" rose high above us, almost exposing her keel to view, and again descended as we in turn rose to the top of the wave, threatening to bury her beneath us whilst the crashing of the breaking upperworks and boats increased the horror of the scene.'

The four-year trip, though, was a complete success. Hooker botanised intensively in Madeira, South Africa, Australia (including Tasmania), New Zealand, Patagonia and the Falkland Islands. He collected and identified over 1,500 species that he then painstakingly documented over the next 20 years, notably in his *Flora Antarctica* (1844–59) and *Handbook of the New Zealand Flora* (1846–67). While travelling, Hooker drew or painted his surroundings and discoveries, including a fish taken from the stomach of a seal, and an active volcano – a sight he said 'surpassed anything that can be imagined'. He later had a sea lion named for him – the now-endangered *Phocarctos hookeri* (Hooker's sea lion). Hooker's habit of meticulously recording the location of plants was the beginnings of his work as a plant geographer, which would become vital to his friend Darwin's theory of evolution.

THE HIMALAYAS

In 1847 Hooker set off again, saying he was intent on 'acquiring a knowledge of exotic botany'. The Scottish botanist Hugh Falconer, who was leaving for India to become Superintendent of Calcutta (Kolkata) Botanic Gardens, and Lord Auckland, First Lord of the Admiralty, independently advised the exploration of the Himalayan region of Sikkim. By taking their advice, Hooker became the first European collector in that region.

On arrival in Calcutta, Hooker was delayed by monsoon rains and political negotiations between the British and the Rajah of Sikkim, so he distracted himself with botanising in Darjeeling, discovering plants such as the ivory-flowered *Rhododendron grande*, the first of his new rhododendrons.

Hooker wrote that he had 'eight thermometers in daily use, 2 barometers, 2 chronometers, a sextant and an artificial horizon'. He also made detailed scientific drawings and painted prolifically, including what is believed to be the earliest Western illustration of Everest. He endured many hardships, including altitude sickness and even arrest. (In November 1849, Hooker and his travelling companion, Archibald Campbell, were arrested by the Dewan of Sikkim in a political manoeuvre against the British government, before eventually being released.) Yet Hooker continued his work assiduously, collecting from 9am to around 5pm every day before returning to his camp to press herbarium specimens and document his collections: 'Large and small moths, cockchafers, glow worms and cockroaches made my tent a Noah's ark by night, when the candle was burning; together with winged ants … flying insects and many beetles, while a very large species of Tipula, swept its long legs across my face as I wrote my journal or plotted off my map.'

A BOTANICAL HERITAGE

Hooker's efforts were rewarded, not least by the 25 new species of rhododendrons he discovered, including the beautiful *Rhododendron ciliatum* (pictured on p116) and many species named after those who had assisted him: *Rhododendron thomsonii* for Thomas Thomson (Scots Doctor, Surgeon in the Bengal Army and Superintendent of Calcutta Botanic Garden); *Rhododendron dalhousiae* named after Lady Dalhousie (wife of the first Marquess of Dalhousie, Governor-General of India); and *Rhododendron griffithianum* (to commemorate William Griffith, British botanist and medic who collected in India and Afghanistan but died before he could publish his observations). Hooker's notes, edited by his father, were published in *The Rhododendrons of the Sikkim-Himalaya* (1849–51), with paintings by the legendary botanical artist Walter Hood Fitch. Sir William Hooker also transformed Capability Brown's 'Hollow Walk' at Kew into a Rhododendron Dell to accommodate Joseph's Sikkim collections; it was described as the Sikkim of Kew and rated by many as the finest display in Britain.

Magnolia cambellii, from Hooker's The Rhododendrons of Sikkim-Himalaya.

In later life Hooker succeeded his father as Director of Kew. He also continued to write scientific papers and floras, including his collaboration with George Bentham, *Genera Plantarum*, containing over one and a half million words, describing all of the known members of 200 plant families, all individually examined. The work took 26 years to complete. On his retirement, Hooker was one of the most influential scientists in the British Empire, continuing his interest in collecting and classifying plants, and botanising in the Lebanon, Morocco and California. He died peacefully at home in Sunningdale, Surrey, aged 94.

Joseph Dalton Hooker:
INSPIRATION FOR GARDENERS

❧ Joseph Hooker was a keen observer of plants. He looked at them carefully and noted all the details of their appearance that could be seen with the naked eye. Close attention to plants is essential to every gardener. Check plants every day and you will notice pests, diseases and nutrient deficiencies before they become established. You will also see if a plant needs staking or watering. Don't just survey your plants overall – look closely.

❧ Hooker was prepared to travel to search for new plants. Finding treasures often requires inspiration, so don't restrict your own plant-collecting expeditions to the local garden centre – be prepared to travel to specialist shows and plant fairs, where you will discover new plants and obtain advice from the experts who grow them.

LEFT *Primula capitata* (round-headed Himalayan primrose) was raised at the Royal Botanic Gardens, Kew, from seeds collected by Hooker in June 1849, during his time in Lachen, Sikkim.

BELOW *Meconopsis napaulensis* (satin poppy) was collected by Hooker at about 3,000m (9,842ft) above sea level in Sikkim, in the Himalayas.

❖ The rhododendrons Hooker collected transformed large country estates with a garden style that re-created natural habitats. Wealthy land owners sought to re-create the Himalayan valleys and forests Hooker described in his journal. Consider the different habitats you can create in your own garden, from ponds and bog gardens to herbaceous borders and alpine troughs. This will provide your own plot with maximum interest.

❖ Hooker kept meticulous records and journals for future reference. Keep a garden book, recording details of your own garden, such as which vegetable varieties are reliable performers and which taste good. You can also note seed sowing dates and keep a plan of borders as a back-up in case labels are lost or moved. Note, too, when plants are fed or harvested. A book of this kind provides an interesting record and is much more reliable than your memory.

❖ Hooker's father, who was Professor of Botany at Glasgow University from 1820 to 1841, noted his son's interest in natural history – and particularly plants – and he nurtured it to fruition. If you have children with a specific interest, do everything you can to encourage their passion. You never know – they may become the next Sir Joseph Dalton Hooker.

TOP RIGHT Hooker named *Rhododendron thomsonii* for his friend Dr Thomas Thomson, who became Superintendent of the Calcutta Botanical Garden.

RIGHT Joseph introduced *Inula hookeri* (Hooker inula) into cultivation from the Sikkim Himalayas in 1849. It's ideal for attracting wildlife; butterflies love the flowers and birds eat the seedheads.

Gregor Mendel

DATE	1822–1884
ORIGIN	SILESIA (MODERN-DAY CZECHIA)
MAJOR ACHIEVEMENT	'EXPERIMENTS IN PLANT HYBRIDISATION'

Gregor Mendel was an Austrian monk who later became abbot of his monastery in Brno. As a result of curiosity, intellect and prolonged experiments using pea plants in his monastery garden, he discovered the basic principles of heredity. Although neither Mendel himself, nor his scientific contemporaries, realised the impact of his findings at the time, his observations were confirmed by later scientists. When his research work was rediscovered in the early twentieth century it became the foundation of modern genetics and the study of heredity. He is now revered as the 'Father of Modern Genetics'.

Pisum sativum, garden pea

J ohann Mendel was born on the family farm on 22 July 1822, in Heinzendorf bei Odrau, a small village in what is now Czechia but in his day was part of the Austrian Empire. As a child Mendel worked in the garden, studied beekeeping and developed a great love and fascination for biological sciences.

When he was 11, his schoolmaster, impressed with his aptitude for learning, recommended he be sent to secondary school to continue his education. The move was a financial strain on his family, and often a difficult experience for Mendel, but despite the challenges he excelled in his studies and graduated with honours in 1840.

Mendel then enrolled in a two-year course at the Philosophical Institute of the University of Olmütz, where yet again he excelled, this time in maths and physics, graduating in 1843. His father expected him to take over the family farm, but instead he joined the Augustinian Order of monks, at St Thomas Abbey in Brno, where he was given the name Gregor.

STUDIES WITH PEA PLANTS

In 1849, when work in the monastery drove him to the point of exhaustion, Mendel was sent to become a temporary teacher in Znaim (now Znojmo in Czechia), which was part of his monasterial duties. However, the following year, Mendel failed the exam for his teaching certificate, so was sent to the University of Vienna to continue his scientific studies. While there he studied mathematics and physics under Christian Doppler (who first described the Doppler effect of wave frequency) and botany under Franz Unger, who had begun using a microscope in his studies and theorised that the cells contained hereditary information, inspiring Mendel to begin his own experiments in this field.

Mendel began experimenting with pea plants (*Pisum sativum*, pictured on p122) in 1854, and continued until 1864, publicising his results in 1855. He selected 34 varieties of pea because they were easy to grow in large numbers, and produced two generations per year, growing tens of thousands of plants in total. Pollination was also easy to manipulate; Mendel either self-pollinated or cross-pollinated the plants himself, ensuring that they were protected from insect pollinators. For his experiments, Mendel selectively cross-pollinated purebred plants with selected traits, then observed the characteristics of their progeny over many generations. The results became the basis for his conclusions about genetic inheritance. Mendel had noticed seven traits in pea plants, which remained consistent over the generations: flower colour (either purple or white), flower position (axil or terminal), stem height (short or tall), seed shape (round or wrinkled), seed colour (yellow or green), pod shape (inflated or constricted) and pod colour (yellow or green). He then cross-pollinated plants with contrasting characteristics so he could examine the characteristics of the offspring. His discovery was that there was never any blending of parental characteristics – one characteristic would simply become dominant and the other would become recessive.

THE LAWS OF INHERITANCE

Mendel delivered two lectures on his theories to the Natural Science Society in Brno on 8 February and 8 March 1865, and the results of his studies were published in 1866 under the title 'Versuche über Plflanzenhybriden' ('Experiments in Plant Hybridisation'). Mendel did not promote his discoveries, so they created little impact in his day, and the few contemporary references to his work reveal that the complex and detailed work he produced was not understood, even by influential people in his field. The importance of variability and its implications on genetics and evolution were usually overlooked, and his findings were not believed to be widely applicable – even by Mendel himself, who wrongly believed that they only applied to certain species or traits.

It was not until 1900, after the rediscovery of his paper, that the results of his experiments were understood. His paper was translated into English by the botanist William Bateson in 1901, and its contents eventually became known as Mendel's Laws of Inheritance. The laws are as follows: the Law of Independent Segregation, which states that inherited characteristics (such as the stem length of Mendel's pea plants) exist in alternative forms, such as tallness and shortness; the Law of Independent Assortment, which states that specific traits operate independently of one another; and the Law of Dominance, which states that for each characteristic, one factor is dominant and appears more often, while the alternative form is recessive – a mathematical formula at a constant ratio of 3:1. For example, when Mendel cross-bred true-breeding purple-flowered peas with white-flowered peas, all of the offspring bloomed purple. When he cross-bred the offspring, three quarters bloomed purple and a quarter bloomed white. Scientists later discovered that Mendel's results not only applied to pea plants but also to most plants, animals and humans.

Many inherited characteristics in *Zea mays* (sweet corn) are controlled by genes that follow Mendel's laws of inheritance – among these are the crop's shape and the colour of its kernels.

Gregor Mendel died on 6 January 1884 at the age of 61, after suffering from kidney problems, and was buried at the monastery; all of the paperwork relating to his research was burned by the abbot, and it would be another 16 years before his results resurfaced. Three scientists all discovered Mendel's principles independently at around the same time – the Dutch botanist and geneticist Hugo de Vries, who had suggested the concept of genes; the Austrian botanist Erich Tschermak; and the German botanist and geneticist Carl Erich Correns. As Correns admitted: 'I thought that I had found something new. But then I convinced myself that the Abbot Gregor Mendel in Brunn, had, during the sixties, not only obtained the same result through extensive experiments with peas, which lasted many years, as did de Vries and I, but had also given exactly the same explanation, as far as that is possible, in 1866.' As a result of their recognition, Mendel, who had the priority of discovery, was posthumously hailed as the 'Father of Modern Genetics'.

Gregor Mendel:
INSPIRATION FOR GARDENERS

❖ Gregor Mendel's crop of choice was peas. In his day, varieties would have grown taller but modern varieties have been selectively bred for mechanical harvesting and growing in gardens. Garden peas are listed according to the timing of their harvest – early, second early (or early maincrop) and maincrop varieties – and some descriptions refer to the pea itself or the pod. Varieties with wrinkled seeds tend to be less hardy and are generally sweeter, although breeding has blurred the boundaries. Petit pois are small and tasty; semi-leafless have more tendrils than leaves, and intertwine as they grow, becoming self-supporting; sugar peas and mangetout are grown for their sweet pods.

BELOW These *Begonia semperflorens* Cultorum Group (wax plant) flowers demonstrate Mendel's Law of Segregation: the red flowers have been inherited from one parent, the white from another.

❖ Mendel's experiments had to span many years, due to the time it takes from sowing to harvest. Earlies mature in around 12 weeks, early maincrops take one or two weeks longer, and maincrops another one or two weeks longer again. Sow in succession for continued cropping and grow some as 'sprouting' seeds for their leaves.

❖ The number of crops Mendel could sow per year was also restricted by the local climate, as peas do not germinate in soil temperatures below 10°C (50°F). Gardeners can start them off in pots under glass or in the greenhouse. Early peas can also be sown in guttering. At planting time, make a trench of a similar profile in the vegetable plot, water the seedlings in the guttering and slide the row into the trench.

❖ Seed companies offer F1 hybrids. Plant breeders cross plants with different desirable features to produce one plant incorporating both. These are again crossed, creating inbred lines of offspring that are true to type, as Mendel discovered. F1 hybrids are usually annuals and vegetable cultivars and are selected for uniformity – usually height of growth, timing of cropping and flavour; they are also larger and more robust, and more able to overcome difficult growing conditions. F1 seed is more expensive to produce; if you save the seed for sowing the following year, they do not come 'true' to the parent. Due to the high cost of maintaining these lines, they are regularly replaced.

TOP If, like Mendel, you want to grow peas, try the easy-to-grow, pretty 'Golden Sweet'. Pick mangetouts regularly to encourage cropping.

ABOVE 'Santonio' F1 hybrid plum tomatoes, an enhanced version of 'Santa' tomatoes, maintain consistency of their characteristics, as predicted by Mendel.

Marianne North

DATE	1830–1890
ORIGIN	ENGLAND
MAJOR ACHIEVEMENT	MARIANNE NORTH GALLERY, KEW

At the age of 40, after spending many years caring for her father, Marianne North decided to realise her dream of travelling the world and painting plants. Between 1871 and 1885 she visited 17 countries on six continents, producing well over 800 botanical paintings. She mostly depicted plants in the context of their natural habitats, providing glimpses of a world that was inaccessible to most, and introducing several new plants to botanists. A gallery at the Royal Botanic Gardens, Kew, houses hundreds of her paintings, which are still in the positions where she placed them.

Amherstia nobilis,
orchid tree

Marianne North was born on 24 October 1830, in Hastings, East Sussex, into a wealthy, cultured, well-connected family; the Norths counted William Holman Hunt, Edward Lear, Charles Darwin, Alfred Wallace, and William and Joseph Hooker among their friends. As a young woman, North received lessons in flower painting, and she became an enthusiastic gardener, often visiting the RHS gardens at Chiswick House and the Royal Botanic Gardens, Kew, where she was given flowers to paint. On one visit, Sir William Hooker gave her the exotic flowerhead of an orchid tree (*Amherstia nobilis*, pictured on p128), the very first to bloom in England. This intensified North's desire, held since childhood, to visit the tropics.

The North family travelled extensively, visiting Europe every summer, and Marianne always took a diary and sketchbook with her. She first used watercolours on a trip to Spain, but it was lessons in oil painting, in 1867, that changed her life. She wrote, 'I have never done anything else since, oil painting being a vice like dram-drinking, almost impossible to leave off once it gets possession of one.' Oils were the ideal medium for the humid conditions she was later to encounter, and one reason why her work has lasted so well.

A TRUE EXPLORER

North had taken over the care of her beloved father after her mother died in 1855, but on a trip to the Alps in 1869 he became ill, and died upon their return. North was devastated, later writing: 'He was from first to last the one idol and friend of my life – apart from him I had little pleasure and no secrets.' She painted to overcome her grief. Aged 40, she sold her Hastings home, devoted herself to botanical art and realised her dream of 'going to some tropical country to paint its peculiar vegetation on the spot in natural abundant luxuriance', as she later wrote. She travelled alone, accompanied only by hired servants to carry her easel and paints.

Between 1871 and 1885, she visited America, Canada, Jamaica, Brazil, Japan, Singapore, Malaysia, Indonesia, Sri Lanka, India, Australia, New Zealand, South Africa, the Seychelles, Chile and Tenerife (where some of the plants she painted still remain at the Sitio Litre orchid garden in Puerto de la Cruz).

In Boston, in the United States, North was overcome by the sight of so many new plants – scarlet lobelias, white orchids and ferns – while in Jamaica, she rented a house in a derelict botanical garden, hanging a bunch of bananas from the ceiling instead of a chandelier, and describing herself as being 'in a state of ecstasy' at the view of palms, orchids and passion flowers from her veranda. Rising at daybreak, she painted *en plein air* until midday, spent rainy afternoons working indoors, and in the evening went out to explore, returning after dark. On her next trip, North visited Brazil, where she wrote that 'every rock bore a botanical collection fit to finish any hothouse in England.' Here she was 'received with much distinction by the Emperor', staying for eight months and completing more than 100 paintings, working from a hut in the rainforest.

Visiting Yosemite on the US West Coast, North witnessed the felling of giant redwoods: 'It broke one's heart to think of man, the civiliser, wasting treasures in a few years to which savages and animals had done no harm for centuries.' Then in a forest on the other side of the world, the Rani of Sarawak described her as having 'skirts kilted up to the knees and heel-less wellington boots, as though born for Borneo jungles'. It was here that North discovered a new species, the *Nepenthes northiana*. This pitcher plant was not the only plant to be named after her; the names of one new genus, *Northea*, and three other species – a feather palm (*Areca northiana*), a red-hot poker (*Kniphofia northiana*) and a poison bulb (*Crinum northianum*) – pay tribute to her achievements. Many species North painted have sadly now disappeared, through what botanist Sir Joseph Hooker described as 'the axe and the forest fire, the plough and the flock'.

Marianne travelled around India for almost 18 months, returning with over 200 paintings. Her diaries ignore the arduous conditions she endured. She wrote that she 'started at four in the afternoon in a big cabin boat ... reached Quilon about 12 the next day ... thence on to Nevereya, where we left the boat and crossed the boundary in a bullock cart. We went on in another canoe, hollowed out of one long tree, for twelve hours more...' She was often forced to pack her paintings while they were still wet, making the final touches on her return to London.

THE GALLERY AT KEW

In 1879, thousands flocked to an exhibition of North's work in London. Later that year, while waiting for a train at Shrewsbury station, she wrote to Joseph Hooker, asking whether he would accept her paintings as a gift to Kew, to hang in a gallery that she would finance. Marianne chose the location, hiring her friend James Fergusson, a well-known architectural historian, as architect, and he designed the gallery in a classical style.

Nepenthes northiana from Sarawak, Malaysia, painted by North around 1876, and later named in her honour by Sir Joseph Dalton Hooker in 1881.

North spent a year designing and painting friezes for the gallery, arranging her paintings – which famously cover the interior like vibrantly coloured wallpaper – and positioning the 246 different types of wood she collected on her travels. When Hooker declined her request to allow refreshments to be served in the gallery, she painted tea and coffee plants over the doors instead.

Shortly after the gallery opened in 1882, as her health began to deteriorate, North travelled to South Africa. Her diaries expressed her frustration that she could not paint faster. After her last trip (to Chile, where she painted the monkey puzzle tree, or *Araucaria araucana*), North added her final paintings to the gallery, which now housed 832 works representing over 900 plant species, ranging in size from a few square inches to a life-size painting of a flower head from Chile, with hundreds of florets. In 1886, she retired to Alderley in Gloucestershire, where she entertained visitors, filled her garden with botanical treasures and died on 30 August 1890, aged 59.

Marianne North:
INSPIRATION FOR GARDENERS

✤ One of Marianne North's paintings portrays a group of North American carnivorous plants, including two pitcher plants – *Darlingtonia californica* and *Sarracenia purpurea* – and the Venus fly trap, *Dionaea muscipula*. All of these can be grown in a shallow tray of rainwater or distilled water, around 1–2cm ($^1/_2$–$^3/_4$in) deep, on a sunny windowsill, or in a conservatory or greenhouse that is frost-free or unheated. Do not feed the plants; put them outside for a few days in summer and let them catch their prey naturally. Remove the flowering spike, as flowers weaken the plant. Towards the end of autumn, when some of the foliage will blacken and die back, keep the compost slightly damp and remove the darkened foliage. They need a cold winter rest, so move plants that are growing indoors to a cooler position like an unheated greenhouse or porch where the minimum temperature is no lower than 0°C (32°F). The plants will come back into growth in late winter. Repot annually in early spring, into carnivorous plant compost in a small 12cm ($4^3/_4$in) pot, and return them to their growing position and their tray of water.

✤ *Darlingtonia* and some species of *Sarracenia* can also be grown outdoors in a bog garden – the best suited for this kind of planting are *S. flava*, *S. purpurea* subsp. *purpurea*, *S.* x *catesbaei* and *S.* x *mitchelliana*.

✤ In Brazil, North painted angel's trumpet flowers (*Brugmansia arborea*) being visited by hummingbirds. *Brugmansia* are large shrubs with hanging bell-like flowers, renowned for their sumptuous fragrance. They are not hardy but make a wonderful feature plant for the greenhouse or conservatory, to stand outdoors in summer. Grow them in John Innes Compost No 3, in a large container in a sunny position. Keep the compost moist and feed with general fertiliser once a month, while in active growth. Water sparingly in winter, maintaining night temperatures of 7–10°C (44–50°F) and day temperatures of 10–12°C (50–53°F). In autumn, cut back the plants so they fit comfortably in the greenhouse, maintaining a height of 1.2–1.8m (4–6ft).

LEFT *Darlingtonia californica* (California pitcher plant), found next to streams and bogs, was among several North American species Marianne painted.

TOP RIGHT North painted *Brugmansia arborea* (angel's trumpet) during her trip to Brazil. It is now classified as extinct in the wild.

RIGHT *Sarracenia purpurea* (common pitcher plant) was included in a painting by Marianne of three North American carnivorous plants.

Luther Burbank

DATE	1849–1926
ORIGIN	UNITED STATES OF AMERICA
MAJOR ACHIEVEMENT	*NEW CREATIONS IN FRUITS AND FLOWERS*

Luther Burbank, the maverick 'Plant Wizard', worked his magic to create hundreds of new plants through rigorous hybridisation and selection. Producing over 800 different kinds of plants during a 55-year career, his creations encompassed fruits, flowers, grains, grasses and vegetables. Not all were successful – there was only one 'pomato', and the 'white blackberry' was a short-lived novelty crop – but one 'sport' of his 'Burbank Seedling', named the Burbank russet, has achieved long-term fame in commercial catering and is now the favoured potato of McDonald's.

Abutilon vitifolium,
Chilean tree mallow

Born on 7 March 1849, on a farm near Lancaster, Massachusetts, the thirteenth of 18 children, Burbank had little more than a high-school education, yet he showed great imagination and creativity from an early age, inventing such objects as a steam whistle made from a willow stem and an old kettle. He also showed great enthusiasm for the plants in his mother's garden.

Early on in life he was inspired by two books: *The Variation of Animals and Plants Under Domestication* by Charles Darwin, and *Gardening for Profit: A Guide to the Successful Cultivation of the Market and Family Garden* by Peter Henderson. On the death of his father, Burbank moved to a small farm in Groton, Connecticut, where he bought 17 acres of land with his inheritance and began breeding plants. He believed that all plants could be improved through selection, hybridisation and grafting, and he selected the best products to breed the best varieties he could.

THE PLANT WIZARD

In 1871 Burbank chanced upon the Early Rose potato and planted 23 seeds. One at maturity proved to be high-yielding, well flavoured and smooth-skinned, with oblong tubers that stored well. Three years later, in 1874 he sold the rights of 'Seedling no 16' – later to become the 'Burbank Seedling' – for 150 dollars (instead of the 500 dollars he'd asked for), kept ten tubers and, soon after, used the money to finance a move to Santa Rosa, 96km (60 miles) north of San Francisco, where three of his brothers had settled, eking out a living from a small nursery. Burbank's fortunes changed and he proved himself a skilled horticulturist, when he took an order for 20,000 Agen plum trees. He germinated almond seedlings in the field, inserting buds of Agen into the growing shoots, and fulfilled the order in nine months. He then sold his nursery to concentrate on plant breeding.

Such was Burbank's brilliance that he created over 800 fruit, flower, vegetable and grain varieties. A showman, who became known as the 'Plant Wizard', he loved 'playing with plants' and pushing the boundaries, creating, among other things, spineless cacti (mainly for cattle food) and the 'plumcot' (by crossing apricots and plums). Some of Burbank's best-known creations include the fire poppy, the Shasta daisy and the 'Burbank July Elberta' peach.

For Burbank it was mass production on an industrial scale. At any one time he would have around 3,000 experiments in progress, involving millions of plants; when working on plums, for example, he tested around 30,000 varieties. Burbank brought plants in from abroad, hybridised, and demonstrating a keen eye, instantly spotting desirable characteristics and making every selection himself.

Burbank wrote several books explaining his work, among them *Luther Burbank: His Methods and Discoveries and Their Practical Application* (1914–15), published in 12 volumes and prepared from his notes of over 100,000 experiments, and *How Plants Are Trained to Work for Man* (1921) in 8 volumes.

SHOWMAN AND INVENTOR

Spurred on by favourable postal rates, Burbank entered the mail-order business, creating his own descriptive catalogue of horticultural wonders. *New Creations in Fruits and Flowers*, published between 1893 and 1901, and the excitement of happy customers, bewitched by his prowess in plant breeding and the many wonderful varieties he created, propelled Burbank to fame and fortune and greater success. By the turn of the century, interest in Burbank was so great that a stream of distinguished personalities visited him in Santa Rosa. These included two other self-made men of genius – Henry Ford and Thomas Edison (with more than 1,000 patents of his own) – who paid Burbank a joint visit on 22 October 1915. Burbank also received the scientist Nikolai Vavilov (see pp170–175) as a guest. (In an obituary, Vavilov would praise Burbank for his creativity, but find weakness in his understanding of genetics.)

In all, along with his famous potato, Burbank introduced 113 plums (many using stock imported from Japan), including his 'stoneless' plum, 10 apples, 16 blackberries (including the Himalayan giant), 13 raspberries, 10 strawberries, 35 fruiting cacti, 10 cherries, 2 figs, 4 grapes, 5 nectarines, 8 peaches, 4 pears, 11 plumcots, 11 quinces, 1 almond, the Wonderberry (which is now back in favour as a 'superfood'), 6 chestnuts,

BELOW 'Rainbow corn' (left) and a selection of plumcots (right) are classic examples of novelty plants raised by Burbank. The latter, a hybrid between a plum and an apricot, has the best flavour when home-grown.

RIGHT Four of Burbank's
extraordinary creations
(clockwise from top left):
the 'iceberg' white
blackberry, the thornless
blackberry, the 'phenomenal
berry' and the 'Macatawa'
blackberry cultivar.

RIGHT Four of Burbank's extraordinary creations (clockwise from top left): the 'iceberg' white blackberry, the thornless blackberry, the 'phenomenal berry' and the 'Macatawa' blackberry cultivar.

3 walnuts, 9 different kinds of grasses, grains and other forages, 26 kinds of vegetables and around 1,000 ornamental plants, from *Abutilon* (such as *Abutilon vitifolium*, pictured on p134) to *Zinnia*, including various species of *Amaryllis*, *Hippeastrum* and *Crinum*, followed by *Lilium, Hemerocallis, Watsonia, Papaver* and *Gladiolus*.

Burbank was often criticised by the scientific community for not keeping meticulous records of his work, but he was more interested in results than in how they were achieved. In 1904, he received a large grant from the Carnegie Institution ($10,000 annually) to promote the scientific study of plant breeding, although this was withdrawn after five years when the reviewer decided that Burbank's methods were more art than science, attributing his success solely to his abilities as a selector.

However, Burbank was undoubtedly a brilliant inventor, a skilful self-publicist in the spirit of P. T. Barnum, and a well-intentioned man who gave generously to others and contributed funds to education, possibly because he felt the lack of an education himself. In 1940 his face appeared on a stamp (in the 'Famous Americans' series) and he was inducted into the National Inventors Hall of Fame in 1986, where he is numbered alongside the Wright Brothers and Steve Jobs.

Burbank died in Santa Rosa on 11 April 1926, and in 1930 the Plant Patent Act in the US was amended to ensure breeders were financially rewarded for their efforts, as an incentive to other growers to continue plant breeding (an early example of plant breeders rights). Edison commented at the time, 'This will, I feel sure, give us many Burbanks'. But there was only one Luther Burbank.

Luther Burbank:
INSPIRATION FOR GARDENERS

❧ Burbank wanted to create the ultimate daisy, his favourite flower. He grew oxeye daisies (*Leucanthemum vulgare*), selected the best, then crossed these with English daisies (probably *Leucanthemum maximum*). He then chose the best hybrids and crossed those with Portuguese oxeye daisies (*Leucanthemum lacustre*), refining these over five or six generations. Deciding that the flowers were not white enough, he cross-pollinated the hybrids with Japanese field daisies (*Nipponanthemum nipponicum*), finally selecting a single plant that had everything he desired – it was long-flowering, with large, white flowers. The Shasta daisy, named for the snow-capped mountain near his home, took 17 years to create and was introduced in 1901. Vigorous and full of flower, it may need staking with brushwood or canes in exposed positions.

BELOW Burbank selected an unusually perfumed *Zantedeschia* (arum lily) and named it 'Fragrance'. Plant arum lilies in moist, rich soil or in water.

❧ The plumcot, a plum–apricot hybrid, was relentlessly selected until Burbank had created a fruit with a plum-like flesh and apricot fragrance. Grow this in a sheltered, sunny position, protected from cold winds, or under glass in colder climates.

❧ One night, Burbank was walking among his arum lilies when he detected the scent of violets. He knew that arum lilies are not fragrant, so he began crawling around in the dark until he discovered the source. He named the selected plant 'Fragrance'. Grow arum lilies in moist soil or in up to 30cm (12in) of water, in sun or part shade.

❧ Burbank spent two decades trying to breed the spines out of a cactus. He wrote, 'For five years or more the cactus blooming season was a period of torment to me both day and night.' He hoped he could transform deserts into productive grazing. Initially his experiments appeared to be a success, but the plants he bred did not like the cold and needed regular watering. Some cacti, such as *Lophophora williamsii* (dumpling cactus) and *Pachycereus schottii* f. *monstrosus*, are naturally spineless and make interesting additions to collections in glasshouses and gardens.

TOP RIGHT *Prunus domestica* (plum) plants, ready to be harvested to make Agen prunes. Burbank called prunes 'educated plums' due to their extraordinary health-giving properties.

RIGHT Burbank wanted to breed a daisy that was longer-flowering, larger and whiter than the rest; the result was the Shasta daisy.

Henry Nicholas Ridley

DATE	1855–1956
ORIGIN	ENGLAND
MAJOR ACHIEVEMENT	*THE FLORA OF THE MALAY PENINSULA*

Henry Ridley was born into a family whose ancestors included
Nicholas Ridley, sixteenth-century Bishop of London, who was burnt
as a heretic; William Penn, founder of Pennsylvania; and John Stuart,
third Earl of Bute, botanical advisor to Princess Augusta, founder of the
Royal Botanic Gardens, Kew. Ridley is remembered for his developments
in tapping rubber trees, his enthusiasm for them as an economic crop,
his intensive botanical collecting and discovery of new species, and his
numerous scientific papers and books on relationships between
animals and plants.

Hevea brasiliensis,
rubber tree

As a boy, Henry Nicholas Ridley was interested in nature, particularly birds and insects, publishing his first paper while at school. On leaving Oxford University he searched for work among the subjects he enjoyed, but after applying and failing to get a zoological post at the Natural History Museum in London, he tried again for a job in the botany department and was successful. Although botany was not high on his list of interests, he threw himself into it with great energy and enthusiasm. While at the museum he published a series of papers on plants from Madagascar, West Africa, Timor and New Guinea, plus two monographs, and in 1887, visited the island of Fernando de Noronha, off the coast of Brazil and published papers on its botany, zoology and geology. In 1888 he was appointed Director of Gardens and Forests in Singapore's Straights Settlements. Travelling at every opportunity, he created a large herbarium of Malayan plants, bringing many to the Singapore Botanic Gardens for further cultivation and study. Ridley also collected information and published a series of reports after a survey of timber trees, rattan palms and other economic plants, rapidly amassing knowledge of the country and its economic products.

RUBBER RIDLEY

While Sir Joseph Hooker (see pp116–121) was Director at Kew, he encouraged the exchange of plants among tropical countries of the empire, highlighting to the India Office the potential importance of many plants, among them, the rubber tree (*Hevea brasiliensis*, pictured on p140). Twenty-two seeds from Henry Wickham's collection in Brazil had been planted by his predecessor, so on his way to Singapore in 1888, on the advice of Hooker, Ridley broke his journey in Ceylon, where there were plantations, to learn more about their cultivation. On arriving in Singapore, Ridley nurtured a plantation of young trees and began tapping experiments on older specimens. His experiments on tapping living trees, without affecting their life span or vigour, revealed an increase in latex flow when a recent cut was reopened, as Ridley had suspected. These discoveries were pivotal to the future success of the industry.

Such was Ridley's zeal for the tree, he spoke to coffee planters at every opportunity, ignoring their lack of interest and urging them to change their crop. His habit of always carrying a *Hevea* fruit in his pocket, ready to talk about it at any moment, gaining him the nickname 'Mad' (and later 'Rubber') Ridley. Convinced of the crop's potential, and sensing the opportunity to establish a rubber industry in Malaya, Ridley continued his experiments with tapping techniques and increased the size of his plantation to be ready with seeds and advice for other plantation holders when that opportunity arose. After the first was planted by Tan Chay Yan in 1896, an epidemic of 'Coffee Rust', which wiped out the Malaysian plantations, was followed by Ridley's predicted rubber boom. Ridley provided a large volume of seeds for the new plantations and shipped them overseas, having improved Wickham's methods of packing. Around the time the plantings matured,

Henry Ford was beginning mass production of cars, which led a growing demand for rubber for tyres. On Ridley's retirement in 1911, he received a gift of £800 from the Malaysian rubber planters, the only financial gain he received for his work. The rubber produced by him was sold to help pay for maintenance of the gardens.

RIDLEY THE BOTANIST

Ridley was also an outstanding field botanist, with a keen eye and vast knowledge. Whenever he travelled, he collected plants. Many species he found were new to science, including 200 orchid species, 73 ginger species and 50 in the genus *Didymocarpus*. He also made several trips to the east coast of Malaya up the Pahang River, and unsuccessfully tried to reach Gunung Tahan, the isolated highest mountain of the peninsula because his food supplies failed, forcing him to return. He also visited Borneo, Sumatra, Christmas and the Keeling Islands, Sarawak and Indonesia. He was also fascinated by plant–insect relationships and wrote on the control of the coconut beetle, pollination of orchids by insects, the habits of the caringa ant, and was the first to report on the white snakes living in the limestone caves of Malaya, and mosquito larvae in pitcher plants.

In 1893, Ridley examined a cross between the Burmese *Papilionanthe teres* (pictured) and the Malayan *P. hookeriana,* describing it in the *Gardeners' Chronicle.*

Ridley produced a great wealth of written material, publishing 38 separate papers and notes in 1906 alone. His *Flora of the Malay Peninsula* in five volumes appeared between 1922 and 1925, and *Dispersal of Plants Throughout the World* in 1930. The final tally of his botanical publications numbers around 350, plus 68 on zoological subjects, 8 on geology and mineralogy, 12 on medical issues, 5 on ethnological matters and 14 biological notes.

Didymocarpus malayanus

Days were not long enough for Ridley to do all that he wanted. Imperfections in his *Flora* were partly due to illness during much of its compilation, and he feared he would not live to complete the work. On his retirement Ridley went to live in Kew, and was publishing new species descriptions in his 91st year. He kept a diary of his travels, his daily life in Singapore and beyond until he lost his sight. He became a Fellow of the Royal Society in 1907, was awarded the Gold Medal of the Linnean Society in 1950, aged 95, making a speech in reply to the president, and from 1892 to 1912 *Curtis's Botanical Magazine* contained at least one of his plants. A 1906 issue, dedicated to him read: 'To Henry Nicholas Ridley, who with untiring generosity has surpassed all recent contributors, enriching the Kew collections with rare and novel plants.' Ridley died at 101 on 24 October 1956, after a monumentally energetic and productive life.

Henry Nicholas Ridley:
INSPIRATION FOR GARDENERS

✤ Henry Ridley found 73 new species in the ginger family, Zingiberaceae. Several in this family make good exotics for cool temperate gardens, being hardier than is often realised, although they dislike waterlogging in winter. Hardy gingers can be grown in large containers and overwintered under glass and encouraged into growth before planting out once the danger of frost has passed. Plant them in an open, hot, sunny position, or in light shade, and improve the soil by incorporating well-rotted organic matter before planting. Protect with a layer of organic matter over winter. Left outdoors, they are often late to appear in spring. *Hedychium densiflorum* is one of the hardiest: 'Assam Orange'

produces spikes of rich orange flowers; *Hedychium greenii* has leaves that are green above and maroon below, with orange flowers later in the growing season; *Hedychium forrestii* grows at altitudes of 200 to 900m (656–2,952ft) and produces beautifully fragrant white flowers.

✤ Although Ridley collected tropical orchids, and native species can naturalise in lawns or areas of grassland, there are also garden-worthy cultivars for shady borders. *Cypripedium* hybrids are ideal for moist, rich, free-draining soil that never dries out, with a pH of 6.5 to 7.0. They must be placed in a sheltered, shady position, away from scorching sunshine. Lighten heavier soils with well-rotted organic matter and sharp sand or grit to improve drainage. Plant in autumn or, if not possible, in early spring, taking care not to damage the roots. Day temperatures should not exceed 30°C (86°F), or fall below 5°C (41°F) in winter. Mulch with pine needles or similar as protection, if needed. Do not dig up plants from the wild. *C. kentuckiense* is dark purple with a creamy yellow pouch; *C.* 'Gisela' is pink-purple with venation on the pouch; *C. reginae* has white flowers and a deep candy pink pouch, and grows well in gardens, coming into growth late in spring. The genera *Pleione*, *Bletilla*, *Epipactis* and *Calanthe* are also suitable for garden borders.

TOP LEFT Ridley collected 73 new species of Zingiberaceae (the ginger family), including *Hedychium forrestii*.

LEFT The richly coloured *Hedychium densiflorum* 'Assam Orange' (ginger lily) was one of the hardiest ginger plants that Ridley collected.

TOP Ridley collected tropical orchids, but *Bletilla striata* (hyacinth orchid) is one of many hardy orchids that would suit a cool, temperate garden.

ABOVE Since Ridley's day, *Cypripedium reginae* (showy lady's slipper orchid) has been extensively hybridised to make it suitable for cooler climates.

Charlotte Wheeler-Cuffe

DATE	1867–1967
ORIGIN	ENGLAND
MAJOR ACHIEVEMENT	MAYMYO BOTANIC GARDEN

A prolific plant collector, painter and letter writer, Charlotte Wheeler-Cuffe travelled extensively in Burma (now Myanmar) with her husband, often to remote, inhospitable places. The first botanist to visit Mount Victoria, she painted plants, found new species and, on the summit of Mount Victoria, discovered what she described as 'a crimson rhododendron brandishing defiance to the four winds of heaven'. With no previous experience she oversaw the design, construction and development of the Maymyo Botanic Garden, proving that strength of character and passion can transform challenges into pleasures.

Rhododendron
cuffeanum

Charlotte Williams (nicknamed 'Shadow' following a serious childhood illness) was born in Wimbledon, London, to a family that had a strong Irish connection – her maternal grandfather was the Reverend Sir Hercules Langrishe, 3rd Baronet of Knocktopher, in County Kilkenny.

Charlotte was taught by a governess at home, studied painting and became adept at 'tinkering' with clocks, French polishing and wood carving. On 3 June 1897 she married Otway Fortescue Wheeler-Cuffe, a civil engineer with the Public Works Department in Burma, whom she had known since childhood. Charlotte confessed in a letter to her aunt, 'I can't help but seeing the comical side of it all! The idea of Otway and me, after jogging along all these years, ending in matrimony'. Two weeks later the couple set off for Burma, from where Charlotte wrote to her aunt: 'I delight in the life out here, it is so full of interest, and the freedom and unconventionality of it suits me down to the ground'. Charlotte created a garden and menagerie there, which at various times included a bear, a leopard and a cockatoo; 'his perch is brought in sometimes as a treat at dinner time.... He loves potatoes & beans and such like, but goes quite silly over milk pudding'!

Charlotte regularly accompanied her husband on official inspection tours, often to remote regions, setting up her easel and painting plants in their native habitat. She produced hundreds of botanical sketches and illustrations, mainly of orchid and rhododendron species, collecting plants as she travelled. She also chronicled her adventures in weekly letters to her mother, and then, after the death of her mother in 1916, to her husband's cousin – over 1,000 in all. Charlotte also corresponded regularly with Sir Fredrick Moore, the Keeper of the Royal Botanic Gardens in Glasnevin, Dublin, exchanging ideas, plants and seeds.

AN EXPEDITION TO MOUNT VICTORIA

In early 1911 Charlotte accepted what she called 'a most tempting invitation' from her friend Winifed Macnabb to spend part of the warm season on Mount Victoria. They were the first botanists to visit the area and to sketch and paint there, and Charlotte returned with a large collection of plants and seeds. She later showed these to the plant explorer George Forrest (see pp152–57), when he was in Burma en route to Yunnan, reporting to Sir Fredrick Moore that, 'Mr. Forrest, a botanical collector who is now in Yunnan tells me that the yellow rhododendron is new.... The white rhododendron grows epiphytically on other trees of any sort, including pines – between 6-8500'....The yellow is a bush & does not grow more than 8' high & flowers like the white one, when quite small'. A parcel duly arrived at Glasnevin containing 12 rhododendrons, 2 orchids, 2 *Selaginella* and 1 fern, along with accompanying paintings. The plant she had labelled 'blue buttercup' was dead, but the seeds were sown carefully, and once they had germinated and flowered, they were identified at Kew as *Anemone obtusiloba* f. *patula*. One of the orchids became *Ione flavescens* (now *Sunipia flavescens*); the yellow-flowered rhododendron was named *Rhododendron burmanicum* and the white-flowered

epiphyte became *Rhododendron cuffeanum* (pictured on p146), in honour of its collector. On a later visit to Glasnevin, Charlotte also met Augustine Henry, an Irish botanist and tree expert who had explored China: 'he & Sir Fredrick Moore waxed quite enthusiastically over my Mt Victoria things & say I must go there again & collect more.'

THE GARDEN AT MAYMYO

On 9 November 1917, Charlotte was visited by the Head of the Forestry Department, Charles Rogers, and William Keith, the Burmese Secretary, who proposed that she design and oversee the construction and initial management of a new botanic garden at Maymyo. She wrote, 'The idea is to have a garden of all the beautiful indigenous flowers, trees & shrubs, with a few imported things ... there are a lot of beautiful things in the area now, including a small patch of primeval forest (rich in native species), a marsh, some rocks ... & weeds.' Charlotte approached the task with relentless zeal, describing it as 'an even more fascinating job than I expected'. Contacts sent seeds from remote areas: rhododendrons, 'a black arum', 'wild flower seeds from the Karind Valley', 'a curious plant from Mansi forest', the coffin juniper (*Juniperus recurva*) *Codonopsis*, bamboo seeds, a small pink *Lilium* (*Nomocharis* species) and three orchids, including *Vanda coerulea*, were among the items from which a garden gradually emerged.

Anemone obtusiloba f. patula was collected by Charlotte at an altitude of 3,800m (12,467ft), from Mount Victoria, Burma, according to a 1915 edition of *Curtis's Botanical Magazine*.

In February 1920, plant collector Reginald Farrer visited the garden, while Charlotte was supervising the installation of the water pipes and metalling of the roads. He later waxed lyrical in *The Gardeners' Chronicle*: 'in a few seasons ... it will rank with Buitenzorg [in Java] and Peradeniya [Sri Lanka; Charlotte later set up a plant exchange between the two gardens] as an object of pilgrimage ... it may even be lovelier ... because a particularly beautiful piece of ground has been chosen ... a winding shallow vale, full of diversities ... and a lake as blue as gentians ... I have no doubt that the Maymyo garden will be paradise.'

In 1921, Otway retired and the Wheeler-Cuffes left Burma. The garden that Charlotte had created was left in the hands of new superintendent R. E. Cooper: 'he is keen & enthusiastic & delightful to work with – so I am handing over my beloved garden to him with great confidence.' Cooper later sent seeds of a beautiful white rambling rose growing in Maymyo to Glasnevin. Its origin was shrouded in mystery for many years until it was found to have been collected by Charlotte, as she wrote, 'at Kutkai on the limestone plateau of the North Shan States ... about 4,300 ft up and near streams'. It was fittingly named *R. laevigata* 'Cooperi'.

After returning to Europe, they settled in the Cuffe family home at Leyrath, outside Kilkenny, Ireland. Shadow, so weak as a child, survived Otway by 33 years and died in Kilkenny on 8 March 1967 – 11 weeks short of her 100th birthday.

Charlotte Wheeler-Cuffe:
INSPIRATION FOR GARDENERS

✤ Charlotte sent 20 Shian lily bulbs (*Lilium sulphureum*) for her mother and family friends and to Glasnevin in exchange for apple scions, informing Sir Fredrick Moore: 'I believe they are not quite as hardy as *Lilium candidum* … there are four here which thrive splendidly in a cold house in winter, and stood out all the summer.' Since the plant hadn't taken well to being repotted, Charlotte recommended renewing a bit of the topsoil instead and earthing up the stems an inch once they have grown 30cm (12in) high. She added, 'You said you had found them difficult, so this may be of interest'.

✤ An Indian horse chestnut (*Aesculus indica*), the only plant left at Glasnevin collected by Charlotte, dates from October 1913. This species is part of a group, including *A. pavia*, *A. flava* and *A. chinensis*, which are resistant to the leaf miner that is currently affecting the horse chestnut. Species from the resistant group should now be planted instead.

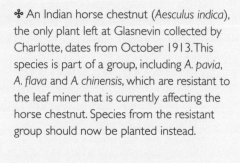

✤ John Besant, successor to Sir Fredrick Moore at Glasnevin, wrote in *The Gardeners' Chronicle* in 1926 about *Anemone obtusiloba* f. *patula*, as it was then in flower blooms for most of the summer. He recommended ordinary soil and a half-shady position for this plant, and also mentioned a white variety called '*Anemone potentilloides* [now *A. obtusiloba* subsp. *potentilloides*], like the last named, introduced by Lady Wheeler-Cuffe. It has not proved so successful in the open, due more to the ravages of slugs than it being tender.'

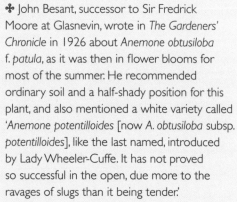

✤ Charlotte said that when approached to design and construct the Maymyo garden, she 'knew nothing about botany. But they insisted and I had to admit I had collected plants all over Burma', adding, 'if it is not well done, I can only plead that I undertook to do my best'. Charlotte undertook the project with great enthusiasm and determination – and succeeded. Never underestimate your potential.

LEFT Contacts sent seeds of *Juniperus recurva* to Charlotte to grow in her botanical garden. This elegant tree thrives in a cool, moist climate.

RIGHT Of all the plants Charlotte originally brought to the National Botanic Gardens of Ireland, Glasnevin, *Aesculus indica* (Indian horse chestnut) is the only one that remains there.

George Forrest

DATE 1873–1932	
ORIGIN SCOTLAND	
MAJOR ACHIEVEMENT PLANT HUNTER AND EXPLORER	

George Forrest was a robust man and lover of the great outdoors – perfect attributes for a career as a plant hunter. During seven expeditions to China over 28 years – mainly to Yunnan, a province rich in flowers – he introduced 1,200 new species. Among them were 509 rhododendrons, over 50 primulas and thousands of seeds, to the excitement of botanists and gardeners alike. Most of his 30,000 herbarium specimens, plus his letters and artefacts, are stored in the archives at the Royal Botanic Garden in Edinburgh.

Primula malacoides,
fairy primrose

As a young man working in a chemist's shop, Forrest produced herbarium specimens, later moving to the herbarium of the Royal Botanic Garden, Edinburgh, where he learned about the classification of plants. He walked 10km (6 miles) to work each day, stood while working, then walked back home again. This idiosyncrasy was noted by Keeper of the Herbarium, Sir Isaac Bayley Balfour, so when Arthur Bulley, a wealthy Liverpudlian cotton broker, wrote asking Balfour to recommend someone to collect plants for his garden from the mountains of Yunnan, Balfour recommended Forrest: 'the head of the herbarium speaks very highly of him. He is a strongly built fellow and seems to be of the right grit for a collector.' Forrest's first expedition amply illustrated that fact.

THE FIRST CHINA EXPEDITION

A year into his time in China, Forrest had been lodging at a Catholic mission, preparing to explore the region near Tibet, when violence broke out; Tibetan lamas had become intent on slaughtering all foreigners in the area. Forrest's group was forced to flee for their lives, and Forrest – having raced down a narrow track and leapt from the path to escape his pursuers, was forced to scramble a few hundred feet down a slope and hide. That first night, in an attempt to escape, he climbed to around 600m (1,968ft) where it was 'very steep, precipitous in parts ... the footing was so precarious, it took me nearly five hours to gain the summit'. When he did, a line of sentries prevented his escape, so he climbed down again and spent the day 'hiding in a hollow under a rock'. The second evening, when he realised the lamas were following his boot prints, he buried his boots, 'descended to the stream, entered the water and waded ... for nearly a mile, taking the utmost care when I got out to leave no track', then carried on barefoot. The journey took all night. On the third night, Forrest ascended a ridge, again blocked by sentries, but found some ears of wheat 'all the food I had for eight days'; 'Once, while lying asleep behind a log in the bed of a stream, I was awakened by a sound of laughing and talking, and on looking up I discovered thirty of them in the act of crossing the stream about fifty yards above my hiding place.'

Throughout this ordeal, it rained heavily. 'By the end of eight days I must have presented a most hideous spectacle, clothes hanging in rags, and covered in mud, almost minus breeches, face and hands scarred and scratched with fighting my way through scrub in the dark, feet ditto, and swollen almost beyond the semblance of feet, shaggy black beard and moustache, and, I have no doubt, a most terrified, hungry and hunted expression on my countenance.' Forrest finally had no choice; he gave himself up at a nearby village with 'only strength to murmur the one word "tsampa" [a local porridge] before I collapsed. Fortunately, I had fallen amongst friends'. His trek to safety was arduous, too. 'Most of the time we had to cut our way through rhododendron and cane brake, and then, when we reached the summit of the range, had two days frightful travelling at an elevation of from 14,000 to 17,000ft over snow, ice, and wind-swept tip-tilted strata, which literally

cut my feet to pieces and shreds and even played havoc with the hardened hoofs of my guides. Bitterly cold it was, sleeping out at such an elevation without covering of any sort. One night it rained so heavily that we had no fire, and had to content ourselves with only a very small quantity of rain water caught in a piece of pine bark. How I scraped through all the hardships I cannot tell.' However, Forrest still managed to take note of the plants he saw along the way: 'The flowers I saw were really magnificent, in fact, so fine were they, that I have decided to run the risk of going back next year if Mr Bulley gives his consent to the arrangement. There was several species of *Meconopsis*, all of them surpassingly lovely, acres of primulas, of which I noted nearly a dozen species in flower, ditto rhododendrons, many of which I had never seen before, and which may probably be new species, besides numberless other flowers.'

Pieris formosa var. *forrestii* was first described from a plant that was raised by seed Forrest collected in the Yunnan mountains, in southwest China.

Despite the trials of his first trip, Forrest returned to China six more times, collecting many iconic plants. Chinese gentian (*Gentiana sino-ornata*) from 'the summit of the Mi Chang Pass between the River Yangtze and the Chungtien Plateau at altitudes between 14-15 000 feet', was granted an RHS Award of Merit in 1915. *Camellia saluenensis* was crossed with *Camellia japonica* by J. C. Williams at Caerhays Castle, Cornwall, to become the first of many *Camellia* × *williamsii* hybrids. When *Rhododendron sinogrande* first flowered in Arisaig, Invernesshire, the garden owner John A. Holmes cut a stem with leaves and flowers, then travelled to Edinburgh and marched towards the Botanic Garden, as later biographer Dr J. MacQueen Cowan put it, 'carrying shoulder high the huge truss, supported by huge leaves, like a magnificent umbrella'. He was stopped, questioned and mobbed before reaching the garden in triumph! The cerise pink-flowered *Rhododendron griersonianum* (pictured right), named for R. C. Grierson Esq. of the Chinese Maritime Customs at Tengyueh, yielded over 150 hybrids, almost one third receiving an AGM or First Class Certificate from the RHS. Among the many primulas Forrest introduced to the world were the hugely popular *Primula malacoides* (pictured on p152) and another that Bulley was so pleased with that he would introduce himself as *Primula bulleyana*.

Rhododendron griersonianum

On 5 January 1932, Forrest died from a heart attack while shooting game. He was buried in a small graveyard at Tengyueh, among the plants and scenery he loved, his legacy as a great plant hunter secured.

George Forrest:
INSPIRATION FOR GARDENERS

✿ *Pieris formosa* var. *forrestii* (pictured on p154), a popular plant for gardens with shade and acidic soils, needs shelter, moist soil and protection from frost when young. According to Forrest, the seeds came from the Tali Range in Yunnan, and the original specimen still exists at Liverpool's Ness Botanic Gardens. Once owned by A. K. Bulley, these gardens are now part of the University of Liverpool.

BELOW *Primula bulleyana* (Bulley's primrose) was grown from seed collected by Forrest and named for his employer, A. K. Bulley. It needs moist soil.

BELOW RIGHT *Abies forrestii* var. *forrestii* (Forrest fir) was discovered by Forrest in Yunnan, China, in 1910. It is fast-growing and sensitive to drought.

✿ When Forrest saw *Magnolia campbellii* subsp. *mollicomata* in the distance in snowdrifts 3m (9³/4ft) deep, he was determined to collect the plant, eventually gathering seeds from it near the Burmese (Myanmar) frontier in 1924. The three seedlings raised were named after the gardens where they were planted. The cultivar 'Lanarth' was originally known by the Williams family who own that garden as 'the Magnolia with the telephone number' after its collection number – 25655 (British phone numbers had five digits at the time). It is highly prized, with a flower colour that was described by G. H. Johnstone in *Asiatic Magnolias in Cultivation* as 'like that of vintage port'. The original plants can still be seen at Lanarth Gardens on Cornwall's Lizard Peninsula.

ABOVE Forrest observed that *Magnolia campbellii* subsp. *mollicomata* took half as long to flower as other forms of *M. campbellii*.

❧ Forrest demonstrated incredible fortitude and resilience, and overcame huge challenges on his journeys. However tough your gardening conditions or unsuccessful your attempts to propagate a plant you should never give up. If you continue, even against the odds, there is always a chance that you will succeed.

❧ Forrest's *Journal of Botany* obituarist wrote: 'It is probable that in quality and richness his herbarium material has never been surpassed. He had developed the technique of plant-drying to a fine point.' The key to his successful collection and preparation of specimens was that he trained local tribespeople to work with him. Having helpers makes a job more pleasant, increases the work rate and brings other skills to a project. Although gardeners are traditionally possessive of their own plot, offers of skilled help are always worth accepting.

ABOVE The autumn-flowering *Gentiana sino-ornata* (showy Chinese gentian), ideal for rock gardens, is one of many special plants introduced by Forrest.

157

Ernest Wilson

DATE	1876–1930
ORIGIN	ENGLAND
MAJOR ACHIEVEMENT	PLANT COLLECTOR

A brown plaque on a terrace house at one end of Chipping Campden
High Street in Gloucestershire marks the birthplace of Ernest Henry Wilson,
'plant hunter in China and the East'; at the other end is the Ernest Wilson
Memorial Garden. Both commemorate a man who introduced more
'garden-worthy' plants into cultivation than any other and is regarded by
many as the most successful plant hunter of all time. Wilson introduced
over 1,000 new plants, took over 800 images on a full-plate camera
and collected over 100,000 herbarium specimens.

Ceratostigma willmottianum,
Chinese plumbago

On leaving school, at 13, Wilson became an apprentice at the nurseries of Messrs Hewitt of Solihull. Three years later, he became a gardener at Birmingham Botanical Gardens, while he studied in the evenings at Birmingham Technical College and won the Queen's Prize for Botany. His next move, aged 21, was to become a student on the diploma course at the Royal Botanic Gardens, Kew, where he won the Hooker Prize for his essay on Coniferae. Wilson was about to enter the Royal College of Science, intent on teaching botany, when he was recommended by the Director of Kew, W. T. Thiselton-Dyer, to Harry Veitch, head of the James Veitch and Sons nursery. Veitch wanted to send a plant hunter to China who would collect seeds of the handkerchief tree (*Davidia involucrata*, pictured on p161). Veitch's objective was clear: 'the object of the journey is to collect a quantity of seeds of a plant the name of which is known to us ... do not dissipate time, energy, or money on anything else.' Herbarium specimens from a tree found in the Yangtze Gorges had been sent to Kew by Augustine Henry, who was working for the Chinese Maritime Customs Service. Veitch, who had an eye for a good garden plant, had noted their beauty.

IN SEARCH OF THE HANDKERCHIEF TREE

Wilson embarked on his first trip on 11 April 1899, travelling via the United States, and stopping briefly at Harvard University's Arnold Arboretum, before arriving in Hong Kong on 3 June 1899. He then headed to Simao, a town in southwest Yunnan, to receive directions from Henry as to the tree's location. Travelling up the Red River, Wilson's boat nearly capsized and lost masts and a sail while negotiating a dangerous rapid. He then joined a mule train to trek 17 days to Simao. Here he finally met Henry, who sketched an area of around 140,000sq km (54,054 sq miles) on a page torn from a notebook, on which he marked where he had discovered the tree, in the 'high mountains ... bordering the upheh and Szechwan provinces'. After a long, arduous journey Wilson 'reached the hamlet of Ma-huang-po and the house where Dr Henry stayed when he found the "Davidia" tree on 17th May 1888'. Did they remember Dr Henry? Yes. Did they know of the tree? Yes. Could someone guide Wilson to the tree? Certainly. After walking for a couple of miles he arrived at the spot, and was confronted by a new house and a *Davidia* stump nearby. 'I did not sleep during the night of 25th April 1900', Wilson wrote in *Aristocrats of the Garden*. Undeterred, he continued 1,500km (932 miles) west to where it was first discovered, and less than a month later (19 May) he 'suddenly happened upon a "Davidia" in full flower! It was about fifty feet tall ... and with its wealth of blossoms, was more beautiful than words can portray'. Wilson collected fruits, too.

AN INTREPID TRAVELLER

After two trips to China, for Veitch, Wilson began collecting for the Arnold Arboretum. On his second expedition, in 1910, Wilson discovered the king's lily (*Lilium regale*) in the arid Min River valley in Sichuan, where he saw them in 'hundreds, in thousands, aye in tens of thousands ... the air in the cool of the

evening is laden with delicious perfume exhaled from every blossom ... its beauty of blossom and richness of fragrance had won my heart and I was determined it should grace the gardens of the Western World'. After returning the following October and gathering the 6,000 bulbs he had previously marked for collection, he was walking along a narrow mountain track, when 'Suddenly I noticed the dog ceased wagging its tail, recoil then rush forward as a small piece of rock hit the path and bounced into the river some 300 feet below'. He shouted to the bearers to put down his sedan chair just as a rock smashed into it, propelling it over the side of the cliff. Another rock then bounced over his head and knocked off his hat. He was just a few yards from safety in the lee of some rocks, when he 'felt as though a hot wire had passed through my leg'. Having been knocked over, he tried to get up but realised his leg was useless; his leg had been broken in two places, and the rock had ripped off his boot and big toenail. Remarkably, Wilson remained conscious and ordered his men to make a splint from his camera tripod. As he lay helpless, a mule train, unable to retrace their route, walked over him, but not one of them stood on him. It was then he 'realised the size of a mule's hoof'. He was carried in a sedan chair to Chengdu, and although infection had set in and amputation looked likely, Dr Davidson, operating for more than an hour, saved and set the leg, leaving him with what he later called his 'Lily Limp'.

Clematis montana var. rubens; this Chinese variety was introduced by Wilson to the nursery of Messrs John Veitch and Sons in 1900.

Wilson made further trips for the Arnold Arboretum (where he became Assistant Director), including one to Japan to collect the Kurume azaleas, and wrote several books, including *Plantae Wilsonianae*, a summary of the woody plants collected in western China for the Arnold Arboretum in 1907, 1908 (when he collected *Ceratostigma willmottianum*, pictured on p158) and 1910. The book was edited by the Director of the Arnold Arboretum, Charles Sprague Sargent.

Davidia involucrata, handkerchief tree

Wilson's life was cut short when his car skidded on wet leaves and crashed into an embankment after visiting his daughter in the United States. His wife died at the scene and Wilson died a week later, on 15 October 1930; both requested to be buried on British soil so were laid to rest at Montreal's Mount Royal Cemetery in Canada.

Ernest Wilson:
INSPIRATION FOR GARDENERS

❖ Wilson found *Clematis armandii* on trees and growing over thickets and bushes. Wilson wrote that this strong-growing, evergreen climber is great for a garden where a 'genial climate prevails' and counted it 'among the most desirable and beautiful of plants I have been privileged to introduce into cultivation.' It has white, fragrant flowers in spring.

❖ *Clematis montana* var. *rubens*, with purple-tinted leaves and vanilla-scented, rose-mauve flowers is hardier than many and, as Wilson wrote, 'acclaimed by garden lovers to be one of the most beautiful of all the clematis'. Flowering from mid- to late June, *Clematis montana* var. *wilsonii* has small white flowers with a fragrance of hot chocolate. Wilson said this was the most common species in west Sichuan.

❖ *Ceratostigma willmottianum* (Chinese plumbago) is an attractive shrub, boasting azure blue flowers from midsummer to early autumn, when the foliage develops red tints (pictured on p158). This is a native of west Sichuan, where Wilson found it in 1908 in the semi-arid regions of the Min River valley. Ellen Willmott raised two plants from seed, sent to her by the Arnold Arboretum; one was grown in her garden

TOP LEFT A native of central China, a *Clematis montana* var. *wilsonii* plant (named for Wilson) first flowered in Britain in July 1909. It produces large white flowers and likes rich, moist, alkaline soil.

LEFT *Kolkwitzia amabilis* (beauty bush) was found by Wilson in Hupeh, China, at about 3,000m (9,842ft), at the watershed of the Han and Yangtze rivers.

at Warley Place in Essex and the other at her sister's garden at Spetchley, Worcestershire. It is best to prune this beautiful plant hard, early in spring, and place it in a sunny spot. It will grow well in chalky soils.

❖ Although *Actinidia chinensis* is commonly known as the kiwi fruit, the alternative name of Chinese gooseberry is more applicable, since seeds were sent from China by Wilson, who collected it in Hubei province: 'The handsomest of all the Actinidias and one of the most beautiful of all climbers.'

❖ Wilson wrote of *Lilium regale* (king's lily), 'The discoverer is fearful lest its admirers undo it with kindness'. He noted that it has certain requirements, and that loam, leaf soil, good drainage and full exposure to sun are absolute essentials. He added, 'Do not give it fertiliser in any form any more than you would give an infant in arms beef-steak.'

BELOW *Actinidia chinensis*, famous for its kiwi fruits (although not as popular as *Actinidia deliciosa*), was collected by Wilson in China in 1900.

BOTTOM *Lilium regale* (king's lily) was collected by Wilson on his first expedition for the Arnold Arboretum in America, just before he and his party were hit by an avalanche.

Francis Kingdon-Ward

DATE	1885–1958
ORIGIN	ENGLAND
MAJOR ACHIEVEMENT	PLANT COLLECTOR, AUTHOR OF 25 BOOKS

The doyen of plant collectors in his day and a noted geographer, Francis Kingdon-Ward devoted his whole life to collecting plants. Over almost 50 years he undertook 22 expeditions, mainly to the Sino-Himalayan region, returning with a host of new species, particularly primulas and rhododendrons. His experiences are recorded in detail in 25 books and innumerable journals, written in his vivid, appealing style. On his last five expeditions Kingdon-Ward was accompanied by his wife, Jean, a resilient and able companion, who is notable as one of an elite group of female plant hunters.

Trollius pumilus, dwarf globeflower (yellow) and *Trollius chinensis*, Chinese globeflower (orange)

Kingdon-Ward was the only son of Harry Marshall Ward, Professor of Botany at Cambridge. Kingdon-Ward went to Christ's College, Cambridge, to study natural sciences but partway through the course his father died, forcing him to finish his degree in two years. Kingdon-Ward then became a junior schoolmaster at a public school in Shanghai. En route in Singapore, he headed out to the rainforest, later writing, 'That was my night out: fireflies and bullfrogs ... I just wanted to steep myself in an atmosphere, to revel in the scents, and to see with my own eyes all the exuberance of life that the warmth, humidity, and equinoctial time-sequence of the tropics produces...'

THE URGE TO TRAVEL

With little enthusiasm for teaching Kindon-Ward joined an American zoological expedition, to Western China, sponsored by the Duke of Bedford, and made a small collection of herbarium specimens. Shortly after his return, Arthur Bulley, who had just lost George Forrest (see pp152–157) to J. C. Williams of Caerhays Castle in Cornwall, was looking for a replacement. Again he wrote to Isaac Bayley Balfour at the Royal Botanic Garden in Edinburgh. 'There is a good man in China now ... with a good knowledge of botany and of plants' was the reply he received. Kingdon-Ward accepted the offer immediately: 'travel had bitten too deeply into my soul and I soon began to feel restless again, so that when after four months of civilised life something better turned up, I accepted it with alacrity', he later recalled. 'Bulley's letter decided my life ... for the next forty-five years.'

Kingdon-Ward's most successful expedition, in 1924, to Bhutan and southeast Tibet, was with Lord Cawdor, where they solved 'the Riddle of the Tsangpo Gorge', one of the deepest gorges in the world. The upper section of the Brahmaputra River, close to Lhasa, is at 3,600m (11,811ft) above sea level, yet where it emerges from the Himalayas and pours into the Assam Valley it is only at a height of 300m (984ft). It had been believed to step down in a series of magnificent cataracts, including the mythical Falls of Brahmaputra, said to be 45m (147ft) high; Kingdon-Ward and Lord Cawdor found it to be a series of smaller falls. On this trip they also collected Tibetan cowslip (*Primula florindae*), dwarf globeflower (*Trollius pumilus*, pictured on p164 with *Trollius chinensis*, another garden-worthy member of the genus), *Berberis tsangpoensis*, several rhododendrons and, among what Kingdon-Ward called 'the wooded hills east of sacred Lhasa', the blue poppy (*Meconopsis betonicifolia*). He wrote, 'among a paradise of primulas the flowers flutter out from amongst the sea-green leaves like blue and gold butterflies'.

A CLOSE SHAVE

Kingdon-Ward faced many challenges on his travels, including battling a fear of heights, suffering bouts of malaria, getting lost in the 'jungle' and surviving by eating rhododendron blossoms, being impaled on a bamboo spike, and escaping when a tree fell on his tent. His greatest escape, however, came on 15 August 1950,

in the Lohit Valley, when he and his second wife, Jean, were caught near the epicentre of the 1950 Assam earthquake, which was 8.6 on the Richter scale, and killed around 4,800 people. Jean recorded the traumatic event in her book *My Hill So Strong* (1952), describing the night when she felt her camp bed give 'a sharp jolt. In a split second it jerked again, more violently and I was dragged roughly from the very brink of sleep back to full consciousness.' She wrote that Francis Kingdon-Ward 'methodically put the cap on his pen' before picking up his lantern and following his wife outside, where they found themselves thrown to the ground, unable to stand or even sit due to the violent shaking of the ground. Jean described the colossal sound they heard as a 'deafening roar that filled all the valley. Mixed with it was a terrifying clatter as though a hundred rods were being drawn over corrugated iron.' As the earthquake continued, Francis, Jean and their porters lay together, face-down in the sand, holding hands as they waited 'in indescribable terror for the enraged earth to open beneath us and swallow us whole.' Miraculously, every member of the group survived this savage ordeal and managed to make their way back to safety.

A LASTING LEGACY

As a botanist, Kingdon-Ward was known for his meticulous scientific approach – his detailed notes were published after each trip by his sponsors. He also had an excellent knowledge of several genera. Between expeditions he worked on identifying plants he had collected, sometimes with other specialists, notably Harry Tagg at Edinburgh's Royal Botanic Garden, an authority on rhododendrons. Kingdon-Ward collected and catalogued over 23,000 plants during his life, and described 119 new species, including 62 rhododendrons, 11 *Meconopsis* and 37 primulas. Several were named after him in later years, including *Rhomboda wardii*, in 1995, from one of his herbarium specimens. In 1956, at the age of 71, he climbed Mount Victoria (3,053m/10,000ft) in southwest Myanmar. A herbarium specimen indicates he collected an *Impatiens*. When it was rediscovered in 2002 in the same location, it was named *Impatiens kingdon-wardii*, because Kingdon-Ward had been the first to collect the species.

Meconopsis betonicifolia (blue poppy) was one of Kindon-Ward's finest introductions; he described its collection in his book, Land of the Blue Poppy.

Kingdon-Ward was honoured with the Founder's Gold Medal and an Honorary Fellowship by the Royal Geographical Society, and received the Victoria Medal of Honour and Veitch Memorial Medal from the RHS. While planning another trip at the age of 72, he had a stroke and died in hospital on 8 April 1958. He was buried in the churchyard at Grantchester, Cambridge, just behind the church. On 10 February 2016, Jean's ashes were buried with him and *Rosa* 'Frank Kingdon Ward' was planted against the churchyard wall in his memory, in the presence of admirers and three generations of the Kingdon-Ward family.

Francis Kingdon-Ward:
INSPIRATION FOR GARDENERS

❧ Kingdon-Ward commissioned Lady Charlotte Wheeler-Cuffe to paint *Meconopsis speciosa* in 1917; the painting was lost and the plant died, too. Not every introduction is successful – even for Frank Kingdon-Ward – but his passion made him keep trying.

❧ Kingdon-Ward wrote of *Meconopsis betonicifolia* (pictured on p166): 'The finest flowers hid themselves … along the banks of the stream. Here among the spiteful spiny thickets of Hippophae, barberry, and rose, grew that lovely poppy'. He noted that several Himalayan poppies do as well in humid climates as in parts of Scotland, Wales and Ireland, but they will not grow just anywhere, and need a good deal of attention.

❧ Kingdon-Ward was adamant about how Tibetan cowslip (*Primula florindae*, named after his first wife) should be grown. He advised that a single plant is 'big and bold' enough to stand alone if its surroundings are big and bold enough, but he felt that *P. florindae* looks its best growing 'in a crowd, as nature intended

BELOW LEFT *Rhododendron wardii* var. *puralbum*, a white-flowered variety of the species named for Kingdon-Ward that flowers in mid- to late spring

BELOW *Rosa* 'Frank Kingdon Ward', bred by Girija and Viru Viraraghavan in India, in 2012. One parent is *Rosa gigantea*, a species collected by Kingdon-Ward and his wife Jean in the mountains of Rajasthan.

it to grow'. He thought that it was the 'forest of stems rising above the cured leaves, like frail masts from a rough green sea which is the attraction; not in a lone specimen.'

❊ After the Second World War, Kingdon-Ward worked for the US government searching for wrecked aircraft over the eastern Himalayas ('The Hump'). During one search he discovered *Lilium mackliniae* (named for his second wife's maiden name) on Mount Sirohi in Manipur. 'That the Manipur lily grows on both sides [of the ridge] – that is to say, in full sun and in half shade – in itself suggests a certain degree of adaptability'. He warned that it will not stand drought or near-drought in the growing season, though, and noted that even slight sunlight bleached the flowers. The plant had grown best on a north-facing slope.

❊ After George Forrest (see pp152–157) met Kingdon-Ward on a ship returning to Southampton, Forrest wrote: 'I must admit much of my prejudice regarding him has gone by the board. He is rather keen we should arrange an expedition together to the country NE of Putao.' Disappointingly nothing materialised, but it is well worth any gardener looking up the huge range of Asian species discovered by these two great botanists.

TOP RIGHT Kingdon-Ward found *Primula florindae* (giant cowslip) growing in the Yarlung Tsangpo basin in northeast Tibet, and named it for his first wife, Florinda Norman-Thompson.

RIGHT Kingdon-Ward discovered *Lilium mackliniae* (Manipur lily) in 1946, in the Manipur region of northeast India, and named it for his second wife, Jean Macklin. It received an RHS Award of Merit when exhibited for the first time in 1948.

Nikolai Vavilov

DATE	1887–1943
ORIGIN	RUSSIA
MAJOR ACHIEVEMENT	*ORIGIN AND GEOGRAPHY OF CULTIVATED PLANTS*

Soviet botanist, geneticist and geographer Nikolai Vavilov was a man of vigour, intellect and charisma, and with one ambition – to end famine in Russia and the world. Knowing this could be achieved by plant breeding and selection, he and his teams combed the world for economic crops and their wild relatives, collecting and preserving genetic diversity. Despite his extraordinary achievements in creating the world's first seed bank and penning a groundbreaking work (*Origin and Geography of Cultivated Plants*) he died an untimely death and became a global martyr for science.

Daucus carota,
wild carrot

As a child, Nikolai Vavilov liked the natural sciences, creating his own herbarium. After studying at the Moscow Agricultural Institute (writing his BA diploma essay on snails as pests) he left Russia and visited the key research laboratories in Europe, meeting many great scientists. Those who influenced him most, with their ideas on genetics, were Rowland Biffen, an agricultural botanist, Director of the Plant Breeding Institute at the University of Cambridge and an authority on wheat rusts and cereal breeding; and William Bateson, Director of the John Innes Horticultural Institution and the first person to use the term 'genetics'. Vavilov also met Charles Darwin's son Francis.

Vavilov realised that the principles of plant geneticist Gregor Mendel and the theories of Darwin could be used in breeding programmes to increase productivity and resistance to drought and disease, but only if the widest possible range of genetic material was available. With this in mind, he and his researchers scoured the world to select as many different kinds of crops and their wild relatives as they could find. These would then be stored in a seed bank in Leningrad for future use.

ESTABLISHING A SEED BANK

Believing that the greatest diversity would be found near the centres of origin, between 1923 and 1931, Vavilov organised, or undertook himself, 115 expeditions to 64 countries over 5 continents. These included Afghanistan, Iran, Spain, Taiwan, Algeria, Eritrea, Bolivia, Korea, Peru and the United States. The goal was to collect seeds on a 'mission for all humanity', creating a seed bank containing over 60,000 seed and plant samples, including around 26,000 varieties of wheat. They not only collected essential crops (wheat, oats, rice and rye) but also forage plants such as clover, fruit, vegetables (like *Daucus carota*, pictured on p170), and plants that produced resins, dyes, textiles and medicines – any plant of use to humankind. On analysis of the data, Vavilov proposed eight Centres of Diversity: I. China, II. India and Indo-Malaya, III. Central Asia, IV. the Near East, V. the Mediterranean, VI. Abyssinia, VII. South Mexico and Central America, and VIII. South America (Peru, Ecuador and Bolivia). The number was modified several times as his research progressed, eventually numbering seven.

The seeds were deposited in the seed bank of the All-Union Institute of Plant Industry in Leningrad, where Vavilov was director. He was honoured for this work with the Lenin Prize, and named as a foreign member of the Royal Society of London and many other venerable institutions. Vavilov had transformed this organisation into a network of 400 experimental stations and laboratories, employing some 20,000 people. His seminal work, a selection of his academic papers entitled *Origin and Geography of Cultivated Plants*, was published in 1926 and translated into English in 1992, revealing his findings. As a result, he became a celebrated academic, and travelled the world, meeting other geneticists including Luther Burbank (see pp134–139).

ENEMY OF THE STATE

Despite his heroic achievements, Vavilov made enemies, notably Stalin's favourite scientist, T. D. Lysenko, once Vavilov's student and 'a young man with green fingers, ruthlessness and boundless ambition', as noted by Vavilov's biographer, Peter Pringle. Famine, drought and Stalin's collectivisation of private farms had led to reduced yields across the Soviet Union. Vavilov, the perfect scapegoat, was held responsible for the famines because his breeding programmes took too long, and the principles of genetics, founded on Western research, were deemed offensive to Communist ideology and branded 'reactionary, bourgeois, and capitalistic', according to Pringle. Lysenko publicly denounced Vavilov at several plant-breeding congresses. Vavilov was then removed from his position, his beliefs on genetics were rejected (and outlawed in 1948) and his reputation was destroyed.

Vavilov was arrested by the NKVD, Stalin's secret police, on 6 August 1940 while on a plant hunting expedition to Ukraine. His 'charges' included sabotaging agriculture, maintaining links with émigrés, belonging to a rightist conspiracy and spying for England. He was subjected to brutal interrogations, then sentenced to death, commuted to 20 years in prison, 'having failed in his duties of applying genetics to practical problems of crop improvement', and for having sent 'worthless expeditions' to gather material for his collection instead of concentrating on local varieties and being unsympathetic to the ideas of Lysenko and Ivan Michurin (a Russian 'Luther Burbank'). The man who had famously stated, 'We shall go into the pyre, we shall burn, but we shall not retreat from our convictions', and had committed his life to eradicating world famine, died in prison of starvation on 26 January 1943. His name was expunged from the Soviet Academy of Sciences' list of members in 1945.

Red flowered *Gossypium arboreum* (Ceylon cotton) and *Gossypium herbaceum* (Levant cotton) – both collected by Vavilov.

Vavilov's staff continued his work at the Institute. Even during the 28-month siege of Leningrad in the Second World War – which resulted in one million casualties – nine members of staff guarding the collections chose to sacrifice their lives and die of starvation, surrounded by thousands of packets of seed, in order to preserve the material for future use, rather than save themselves.

Vavilov was posthumously pardoned, absolved of his 'crimes' and 'rehabilitated' after Stalin's death in 1953. The organisation he worked for was renamed the N. I. Vavilov Research Institute of Plant Industry and awarded the Order of Lenin in 1967 and the Order of Friendship of Peoples for its scientific achievements. After Stalin's death, hundreds of books and articles on Vavilov's life and accomplishments were published and memorial displays were opened in Moscow, St Petersburg and Saratov, where he died. The Vavilov Institute in St Petersburg is now one of the largest and most varied stores of plant germplasm in the world.

Vavilov was often quoted as saying, 'Time is short, and there is so much to do. One must hurry.' When it comes to conserving genetic diversity, the work he started remains a priority today.

Nikolai Vavilov:
INSPIRATION FOR GARDENERS

❖ Vavilov recognised that 'genetic erosion' endangered human survival. If we lose plants, in the wild or old varieties in gardens, what will be left for future breeding? Gardeners can maintain genetic diversity by growing heritage varieties and supporting 'seed saver' groups like the Irish Seed Savers Association, the Seed Savers Exchange in the US, Seed Savers in Australia, and Garden Organic's Heritage Seed Library and Seedy Sunday events, both in Britain. Old ornamental varieties need saving, too, through organisations like Plant Heritage in the UK.

❖ Create your own collections of 'heritage' varieties to preserve the genetic diversity of your area. They may be apples from the county where you were born, or the county where you now reside – or possibly apples recommended by others, such as specialist local nurseries, or by the notable Edwardian pomologist and gastronome Edward Bunyard.

❖ Heritage varieties are usually selected for taste and suitability to local conditions rather than their ability to be transported great distances to supermarkets. You will find them very tasty but not always highly productive – they do not always store well.

❖ Scientists have bred disease resistance into garden crops to reduce the need for chemicals. Some of the most disease-resistant potatoes, for example, are the Sarpo varieties bred in Hungary – particularly Axona, which has excellent resistance to potato blight. However, resistance can always be overcome by a massive attack.

❖ The Svalbard Global Seed Vault opened on 26 February 2008. It is housed in an old copper mine, 120m (394ft) inside a mountain in Norway, and has enough space for four and a half million seed samples; seeds can be stored for future use, from 2,000 to 20,000 years. The Seed Vault contains duplicate specimens of genetic strains from seed banks around the world, including material collected by Vavilov. The N.I. Vavilov Research Institute of Plant Industry has deposited 5,278 seed samples of 112 species originating from 96 different countries in the vault. In a domestic setting, if you are saving seed for use next year, use an airtight labelled container in the fridge, together with a packet of calcium oxalate crystals to control humidity.

LEFT Vavilov and his researchers collected *Avena sativa* (cultivated oat); it is widely used by humans but most commonly eaten by livestock.

TOP Vavilov thought that *Prunus armeniaca* (apricot) originated in China. Cultivars have now been bred to tolerate various climates.

ABOVE *Colocasia antiquorum* (Arabian cocoa root) is a staple crop in the tropics. Vavilov identified central and western China as its potential centre of origin.

Archibald C. Budd

DATE 1889–1960	
ORIGIN ENGLAND	
MAJOR ACHIEVEMENT *FLORA OF THE CANADIAN PRAIRIE PROVINCES*	

Although he had no formal training, Budd, a 'humble student of the wonders of nature' and an avid reader and astute observer of plants, transformed himself into a highly respected self-taught botanist and a keen student of taxonomy. After emigrating from England to Canada – a move financed by winning a limerick competition – he began working at an experimental station where he gained a knowledge of agricultural weeds. He ultimately became an authority on the flora of the prairies and shared his knowledge and enthusiasm through lectures, numerous articles and his greatest achievement, his *Flora of the Canadian Prairie Provinces*.

Axyris amaranthoides,
Russian pigweed

Archibald Charles Budd was born in London on 28 April 1889. He attended Bellenden Road Higher Grade School in Peckham where he achieved a merit certificate. As a child he was fascinated by natural history, particularly aquatic life and insects.

He began his working life in the British Civil Service at the Customs House and later for the General Post Office. However, in 1910 his future was transformed when he won £250 as the first prize in a limerick competition (roughly equivalent to £28,750 in 2024). This sudden fortune enabled the 21-year-old to move to Canada, where he first worked as a farmhand at Waldek, then acquired a homestead south of the small village of Rush Lake in Saskatchewan Province, in the west of the country. Realising he had no knowledge of the flora on which to feed some caterpillars he had collected, Budd invested a dollar in the *Farm Weeds* bulletin. This small purchase turned his interest in plants into a lifelong passion and study.

His knowledge gained a practical use in 1926 when he began working at the Dominion Experimental Station at Swift Current, Saskatchewan. The station had been established six years previously, one of several whose role was to improve the effectiveness of local agriculture, in this case the agricultural methods for the dry areas of southern Saskatchewan and Alberta. Here Budd's career progressed from gardener-labourer to soils researcher, weed taxonomist (assisting in the naming and classification of plants), and studying the physiology of weed seeds.

From 1944 until his retirement in 1957 Budd was a Range Botanist, making recommendations for cattle stocking rates, timing of grazing and identification of beneficial or detrimental plant species. The literature he published indicates he worked on the identification and biology of rangeland species as well as some introduced, problematic agricultural weeds alongside other native flora. A quick learner, he shared his knowledge and enthusiasm freely, enabling many fellow professionals to achieve a similar high level of learning so they could appraise rangeland and carry out ecological studies themselves.

PRESERVING SPECIMENS

In 1932, after collecting and pressing plants for a display of mounted specimens at the World Grain Show in Regina, Budd found a new interest in preserving specimens. This not only became a hobby but, combined with his knowledge of plants, led him to become Curator of the herbarium at the Experimental Station where he added over 7,000 plants; he also assisted in establishing an herbarium at the Saskatchewan Museum of Natural History, one of the finest collections in Western Canada at the time. He later donated his personal collection to Swift Current Museum. The National Collection of Vascular Plants of Agriculture and Agri-Food Canada houses two specimens of *Ranunculus glaberrimus* and one specimen of *Verbascum thapsus* that were collected by Archie Budd. His curiosity for plants knew no bounds, he was fascinated with everything from pestilent agricultural weeds to the most beautiful wild flowers.

FLORA OF THE PRAIRIES

In 1949, after years of research, he produced a simple key to the *Flora of The Farming and Ranching Areas of the Canadian Prairies*, an accessible reference work aimed at amateur field botanists, agricultural representatives, farmers and ranchers who needed to identify plants but were not trained botanists. The publication was enthusiastically received by both amateurs and professionals alike. He consulted existing floras and fellow professionals, but his main source of information was the herbarium at the Swift Current Experimental Station.

The work's success led to a revision and a reprint of 5,000 copies in 1952 by the Canadian Department of Agriculture; it was then revised and expanded many times before finally becoming *Wild Plants of the Canadian Prairies* in 1957, the year of Budd's retirement. Regarded as a definitive work on the subject, and expanded to include areas of the Rocky Mountains and boreal forests, Budd assisted with further revisions until his death in 1960. It was later enlarged to 872 pages in 1987 by his colleagues J. Looman and K. F. Best, who renamed it *Budd's Flora of the Canadian Provinces - A Field Guide to the Plants of Alberta, Manitoba and Saskatchewan* in his honour. It remains a definitive work on the subject. Renowned for its thoroughness and ease of use, it includes family, genus and species, followed by concise descriptions of each species plus details of habitat and vegetation zones in the region. It even includes short and midgrass prairies and niche habitats within them, such as eroded areas which are home to rare cushion plants and the three native cacti, noting that *Opuntia polyacantha* a type of prickly pear was 'Very common; on dry prairies and light soils, increasing greatly through overgrazing and erosion as human impact increased.'

Opuntia polyacantha, a prickly pear, is native from central Canada, south to northeast Mexico.

The text is supplemented by Budd's exquisite, detailed drawings and photographs of his carefully pressed specimens. It is an astonishing *magnum opus*, particularly when you consider the fact that it was originally the work of a single, dedicated enthusiast.

Budd also shared his knowledge in *Blue Jay*, the journal of the Saskatchewan Natural History Society, writing or co-authoring on a range of subjects from 'The tall docks of Saskatchewan' to 'Still More Cypress Hills Plants' an article recording two plants, the romboid leaved saxifrage (*Saxifraga rhomboidea*) and the white hawkweed (*Hieracium albiflorum*) – both species from the Eastern Rockies, whose only location in the Saskatchewan region was the Cypress Hills.

Budd was an enthusiastic and knowledgeable lecturer, whether describing spring emerging as the snow melted from the hills or the minute details of individual species, notably the Russian pigweed (*Axyris amaranthoides*, pictured on p176), an invasive crop weed. The plants *Ranunculus glaberrimus* var. *buddii* f. *monochlamydeous* and *Ranunculus glaberrimus* var. *buddii* were named to commemorate Budd; unfortunately, they are not legitimate as they were never formally published. However, *Budd's Flora* remains testament to how talent, enthusiasm, self-education and hard work resulted in a publication that remains of long-term benefit to many.

Archibald C. Budd:
INSPIRATION FOR GARDENERS

❖ Writing on *Thuja plicata* (Western red cedar), Budd notes, 'leaves not distinctly keeled and resin gland inconspicuous. Moist woods; southern Rocky Mountains'. This makes a large, pyramidal specimen tree with drooping branchlets but is more often used as a fast-growing evergreen hedge. It tolerates anything but dry conditions, prefers moist, well-drained soil in sun and grows better in wetter climates. The decay-resistant wood is used in garden buildings.

❖ There are notes and a photograph of an herbarium specimen of *Symphyotrichum laeve* (formerly known as *Aster laevis*) in *Budd's Flora*. The smooth aster is mildew resistant and important in the breeding of many *Aster novi-belgii* cultivars. It forms robust clumps of lance-shaped leaves with lavender-shaped ray florets.

❖ *Camassia quamash* is an upright perennial bulb reaching around 30cm (12in) tall, which produces deep blue flowers between the conclusion of spring flowering bulbs and before herbaceous perennials appear. Flourishing in sun or part shade and heavy moist soils that can be dry in summer, it is ideal for planting in meadows. Its flowers are attractive to bees and avoided by slugs and snails. The edible bulbs can be roasted or boiled – they were widely eaten by First Nations peoples of North America. *Budd's Flora* says they are found in 'moist meadows in the Southern Rocky Mountains'.

❧ *Cornus canadensis*, also known as *Chamaepericlymenum canadense*, is recorded in *Budd's Flora* as 'very common; in shady woodlands; throughout' and was described in an article he wrote for *Blue Jay* in spring 1953. This low-growing ground cover needs cool conditions at altitude or in northern climes where temperatures are no higher than 18°C (64°F). It has pretty, white bracts, surrounding tiny greenish flowers which are followed by clusters of small red fruits. It is a miniature flowering dogwood.

❧ *Budd's Flora* records *Aquilegia canadensis*, the wild columbine, as being found 'in open woodlands; eastern Boreal forest, reported from Qu'Appelle Valley in Saskatchewan.' This stout, upright herbaceous perennial, growing up to 80cm (31¹/₂in) tall, prefers moist soil and sunshine and produces gorgeous large, nodding, scarlet and lemon-yellow flowers. Short lived, it self-seeds freely and if isolated remains true to type.

LEFT *Thuja plicata* was discovered by settlers during the Malaspina expedition (1789–1794) and described from a specimen collected by Nootka Sound by Taddaeus Haenkel, assistant to Née, chief botanist on that expedition.

TOP RIGHT *Camassia quamash* is just one of several garden-worthy species in this attractive genus, generally with vivid blue flowers. *Camassia leichtlinii* subsp. *leichtlinii* is white flowered; its buds appear when the blue forms of the plant are flowering.

RIGHT *Cornus canadensis*, a low-growing cornel or dogwood, comes from a genus ranging from creeping groundcover to shrubs and trees, grown for their coloured stems or ornamental bracts with insignificant individual flowers in the centre.

E. K. Janaki Ammal

DATE	1897–1984
ORIGIN	INDIA
MAJOR ACHIEVEMENT	PIONEERING CYTOLOGIST AND PLANT BREEDER

Janaki's achievements were astonishing for a woman of her era. Born into a large family, she was highly educated at a time when women were expected to stay at home, went abroad to study and work when it was taboo for Indian men or women to do so, and became a leading cytologist and plant breeder. Despite her focus, the 'woman who sweetened India's sugar cane' also retained an interest in ethnobotany, plant geography and 'all plants', remaining humble and dedicated to her life of science until the end.

Nicandra physalodes,
shoo-fly plant

Geeta Doctor once described her Aunt Janaki as a 'tall and commanding presence in her prime. She tied her lustrous long hair into a loose bun at the nape of her neck. In her later years, she took to wearing brilliant yellow silk sarees with a long loose blouse or jacket in the same colour. Her statuesque presence reminded people of a Buddhist lady monk ... she took a vow of chastity, austerity and silence for herself, limiting her needs to the barest minimum.'

Janaki Ammal was born in Tellichery, Kerala, on 4 November 1897, the tenth of 19 children of Dewan Bahadur E. K. Krishnan, a sub-judge. Her father corresponded regularly with academics, maintained descriptive notes about his developing garden, took a keen interest in the natural sciences and wrote two books on birds. The family was cultured and open-minded and Janaki was encouraged to engage in intellectual pursuits at a time when most Indian girls did not even attend school, let alone develop a career.

Ammal displayed an early passion for botany and fulfilled her potential with extraordinary speed. On leaving school, she moved to Madras, obtained a BA from Queen Mary's College, and then a BSc (Hons) from the Presidency College in Chennai in 1921, where she became interested in cytogenetics. She then won the Barbour Scholarship from the University of Michigan, where she gained her master's degree and became the first woman to obtain a PhD in botany in the United States – for her 1931 thesis, 'Chromosome Studies in *Nicandra physalodes*' (the shoo-fly plant, the focus of this paper, is pictured on p182).

A CAREER ABROAD

Between 1934 and 1939, Ammal worked as a geneticist at the Sugarcane Breeding Institute at Coimbatore in Tamil Nadu, founded to develop sweeter hybrids for the Indian sugarcane industry. Her research played a vital role in selecting the best varieties for a breeding programme. She succeeded in creating several high-yielding strains that thrived in Indian conditions, reducing the need for imports and turning India into a major sugar-cane producer. She also identified *Saccharum spontaneum* as an Indian variety and produced hybrids between related plants, and was the first to cross sugar cane with other members of the grass family like *Zea*, *Sorghum*, *Imperata* (including *Imperata cylindrica*, pictured on p185) and bamboos.

Yet despite her brilliance, Ammal's status as a single female scientist created problems among her male peers, so she left for England just after the beginning of the Second World War and became Assistant Cytologist at the John Innes Horticultural Institution in London. She would often dive under her bed during night-time bombing raids, returning to work the following day to brush broken glass from the shelves and continue her research. Her studies on chromosome numbers traced the evolution of many species and varieties, and she detailed much of this work in her *Chromosome Atlas of Cultivated Plants*, written with C. D. Darlington, Director of the Institution. Her staff file also noted 'she smuggled a Palm Squirrel into the country and it was kept at J.I.I. for many years. Its name was "Kapok".'

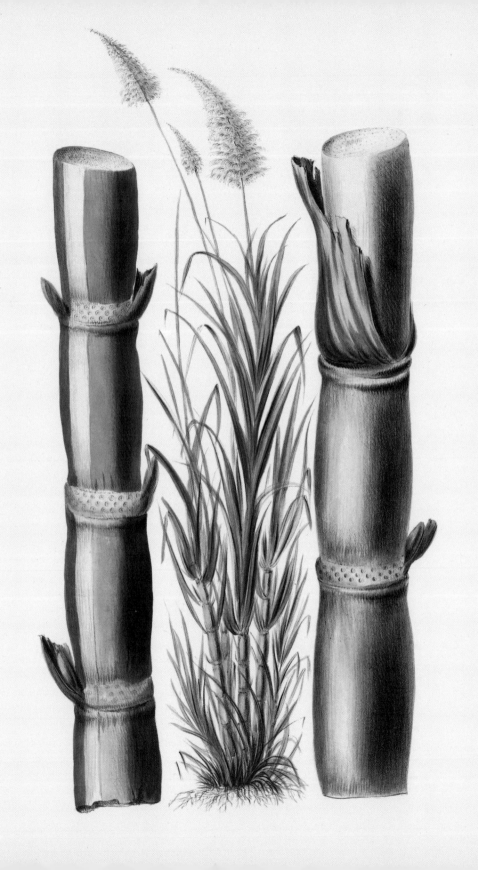

Ammal then became Cytologist at the Royal Horticultural Society at Wisley, where she analysed chromosomes of a wide range of garden plants, lectured, created new hybrids and answered questions from gardeners. She also authored several articles in the *Journal of the Royal Horticultural Society*, reflecting her interests, including 'Chromosomes and the Evolution of Garden Philadelphus', 'Chromosomes and Horticulture', 'The Chromosome History of the Cultivated Nerines' and 'The Story of Primula Malacoides'. Ammal examined a Plymouth strawberry plant sent by plantsman E. A. Bowles during her research on the origins of cultivated strawberries and corresponded with him about the identity of a crocus. On 7 November 1949 she wrote: 'I am glad to confirm from chromosome counts your doubts about the Crocus distributed as *C.karduchorum* being only an unspotted variety of *C.zonatus*. The bulb you sent was 2n=8 ... the true *C.karduchorum* is 2n=20.'

Ammal also investigated the effects of colchicine (a chemical found in autumn crocus, or *Colchicum autumnale*) on the cell division of a number of woody plants. Applying it in solution to the growth tip of seedlings once the seed leaves have expanded doubles the number of chromosomes. Ammal treated several *Magnolia kobus* seedlings and one was selected and named after her. 'Janaki Ammal' is a vigorous, multi-stemmed tree, over 6m (19½ft) tall and wide, producing masses of white flowers, with varying numbers of petals, over several weeks, and heavily textured leaves. The original specimen can still be admired on Battleston Hill at Wisley. In 1951 Ammal returned to India at the invitation of the Prime Minister, to reorganise the Botanical Survey of India (BSI), eventually becoming Director General.

Ammal experimented by hybridising one of the world's major crops, *Saccharum officinarum* (sugar cane), with other genera and species.

Not only was Ammal a brilliant cytologist, she was also a great thinker. She proposed that the Chinese and Malaysian influence on the flora of northeast India had led to natural hybridisation between them and the native Indian plants, leading to greater plant diversification, while she attributed the higher rate of plant speciation in the cold and humid northeast Himalayas, compared to the cold and dry northwest, to polyploidy, creating greater variation in plants.

Imperata cylindrica, cogon grass

Even after retirement, Ammal continued her research at the Centre for Advanced Study in Botany field laboratory, creating a garden of medicinal plants and rearing of a large family of cats to study their subtle differences. She died at work of natural causes on 7 February 1984.

Ammal was a woman of humility and an early follower of Gandhi, who even when Director of the BSI was often seen sweeping the road outside. She refused to speak about her career, saying: 'My work is what will survive.' It remains a lasting testament to one of the most remarkable cytologists and women of her generation.

E. K. Janaki Ammal:
INSPIRATION FOR GARDENERS

✤ In an article entitled 'The Origin of the Black Mulberry' for the *Journal of the Royal Horticultural Society*, Ammal wrote: 'The black mulberry has the highest chromosome count of any flowering plant … the origin of this species therefore presents a problem of special interest.' She noted that this plant had not been recorded in the wild and had been a cultivated plant from 'very ancient times'. She linked the development of the species with its increasing chromosome numbers. Long-lived and often seen as very old trees in gardens, there are also dwarf varieties of the black mulberry plant, such as *Morus* Charlotte Russe 'Matsunaga', which are ideal options for containers and small gardens, and are also highly productive.

✤ Ammal also studied triploid *Kniphofia snowdenii* (now *Kniphofia thompsonii* var. *snowdenii*), taken from Colonel F. C. Stern's garden at Highdown, which she found to have more chromosomes than the other species. It had been discovered by Mr Snowden of the Department of Agriculture, growing in short grass and scrub at 2,400–3,000m (7,874–9,842ft) up on Mt Elgon, Uganda, in 1916. Knowledge of a plant's native habitat allows gardeners to put the right plant in the right place, so that it will flourish. This approach

BELOW Ammal wrote about the genetics of *Morus nigra* (black mulberry), which is noted for its large number of chromosomes.

was first presented by Andrew Chatto, and was the basis for the choice of plants at Beth Chatto's garden. It is a recommended principle for your own garden, too.

✤ In 'Chromosomes and Horticulture', Ammal used the strawberry as an example, and quoted from Shakespeare's *Richard III*, when the Duke of Gloucester urges his host, the Bishop of Ely, to send for some home-grown strawberries. These would have been *Fragaria vesca* (woodland strawberries). When Sir Isaac Newton ate strawberries, he would have eaten *Fragaria elatior* (musk strawberries), introduced from Europe, before his death. *F. chiloensis* (beach strawberries) and *F. virginiana* (scarlet strawberries) were introduced from the New World. Fifty years later these were hybridised to produce 'Royal Sovereign', a regular Chelsea gold-medal winner for the great gardener Beatrix Havergal, and the strawberries of today.

TOP *Kniphofia thomsonii* var. *thomsonii* (Thomson's red-hot poker), grown by great gardener Fredrick Stern, was analysed in detail by Ammal.

ABOVE Ammal described the origins of garden strawberries using *Fragaria virginiana* (scarlet strawberries). These were first reported in 1766.

William Stearn

DATE 1911–2001	
ORIGIN ENGLAND	
MAJOR ACHIEVEMENT *DICTIONARY OF PLANT NAMES FOR GARDENERS*	

Professor William Stearn, a polymath of immense academic brilliance, was a kindly, modest man. Denied a university education, he was almost entirely self-educated and published his first scientific paper while working as a shop assistant. During a career based first at the library of the Royal Horticultural Society, then later at the Natural History Museum, he wrote around 500 books, articles and monographs, ranging from detailed studies of plant genera, through histories and biographies, to books on botanical taxonomy, many of them becoming standard works of reference for both botanists and gardeners.

Hosta plantaginea,
August lily

W illiam Thomas Stearn was born in Chesterton, Cambridge, on 16 April 1911, the eldest of four sons of a coachman. He gained a scholarship to Cambridgeshire High School for Boys, where he was inspired by an outstanding biology teacher and became secretary of the Natural History Society. While still at school, Stearn attended an evening course in palaeobotany taught by Professor of Botany A. C. Seward, who recognised his potential and gave him access to the herbarium and library at the University of Cambridge Botany School, where the young man spent most of his evenings and weekends. John Gilmour, curator of the herbarium there, soon realised the extraordinary ability and enthusiasm of his assistant.

Stearn's father had died when Stearn was 11, and his widowed mother lacked the means to support him through university, so he became a shop assistant and cataloguer at Bowes and Bowes, university booksellers in Cambridge. It was during this time that he published his first scientific paper in 1929, after finding a campanula with a distorted corolla in a garden – 'A new disease of *Campanula pusilla* (*Peronospora corollae*)' – thereby documenting its first known appearance. This was followed by several papers on several genera of garden plants, among them *Poncirus*, *Mimulus* and *Vinca* – 24 papers in all – followed several years later by his first major monograph on *Epimedium* and *Vancouveria*, which he worked on while studying the genus *Allium*, writing 'A Bibliography of the Books and Contributions to Periodicals Written by Reginald Farrer', and making a significant contribution to the catalogue of the Cambridge herbarium.

THE MOVE TO LONDON

In 1933, after a recommendation to the Royal Horticultural Society by the gardener E. A. Bowles, Stearn was made assistant librarian at the Lindley Library in London; he was later promoted to Chief Librarian. While there, Stearn began to teach himself Swedish, the language of the eighteenth-century botanist Carl Linnaeus (see pp56–61), and started to travel to botanical centres throughout Europe. He served in the RAF Medical Service during the Second World War, then returned to the RHS, moving into the library while searching for somewhere to live.

Just before the war, the Director of Wisley, Frederick Chittenden, had become Editor-in-Chief of a new *RHS Dictionary of Gardening*; when war broke out and many botanists were called up for active service, Chittenden ended up writing much of the text himself. When he died in 1950, Stearn completed the task, compiling entries on more than 500 plants, writing the fourth volume in just six months and becoming, as he often recalled, 'a peculiar authority on plants from "So-" onwards'. In 1950, he also co-authored *The Art of Botanical Illustration* with Wilfrid Blunt, the standard authority on the subject.

After attending the International Botanical Congress in Stockholm in 1950, Stearn drafted the first edition of the *International Code of Nomenclature for Cultivated Plants*, governing the naming of garden plants. This was published in 1952, the

same year that Stearn moved to the Natural History Museum, where he eventually became Senior Principle Scientific Officer. He remained there until his retirement in 1976, and around 200 publications date from this period.

In 1976 Stearn was awarded the Gold Medal of the Linnean Society for his 176-page introduction to the facsimile of Linnaeus's *Species Plantarum* (1957). It was later written in his *Daily Telegraph* obituary that 'No other botanist had the knowledge of botanical history and the linguistic ability in Latin, German, Dutch and Swedish needed to write what is regarded as a classic study of the great naturalist'. Much the same was said of another of his outstanding works, *Botanical Latin: History, Grammar, Syntax, Terminology and Vocabulary* (1966), which could only have been written by a unique individual with a diverse knowledge of botany, taxonomy and the construction of Greek and Latin; it was a book that many thought could never be written. Although designed to help students, it became a standard work for plant taxonomists and remains so to this day.

LATER LIFE

Visits to the Cambridge Botanic Gardens, gardening at home and his work with the Royal Horticultural Society gave Stearn a lifelong interest in horticulture, amply reflected in his *Dictionary of Plant Names for Gardeners* (1992), revealing the fascinating stories behind the Latin names of popular plants (including *Hosta plantaginea*, pictured on p188 – *plantaginea* meaning 'resembling a plantain').

Stearn wrote about *Poncirus trifoliata* (Japanese bitter orange), the only member of its genus. It is a fragrant but viciously spiny plant.

After retirement Stearn continued his research at the herbarium and library of the Royal Botanic Gardens near to his home at Kew, in his final years, returning to the genus *Epimedium*, the subject of his first monograph, examining specimens in his hospital bed, many of them new to science, until shortly before he died.

Stearn received many honours and accolades but the one that gave him most pleasure was his first honorary doctorate from Leiden University in 1960. He also became an honorary professor of Reading University and a visiting professor at Cambridge University's Department of Botany, and received the RHS's Victoria Medal of Honour in 1965.

At one time Stearn's entry in *Who's Who* listed his recreations as 'gardening, walking', until his family changed it to 'gardening, talking'. Dr Max Walters wrote in his obituary in the *Guardian* in 2001: 'I am not the only inquirer to have benefited from his affable, patient care, which was sometimes positively uncanny. I recall an international botanical gathering in Cambridge some 20 years ago; we had set out in a coach for some local nature reserve, and I took the opportunity of asking William about cultivated varieties of cannabis, the illegal growing of which was just becoming a problem. He answered my question adequately, and we proceeded on our outing. But, on the return journey – before any of us had been back to Cambridge – he presented me with a copy of his [scientific] cannabis paper, which he "thought I might like to have". I never learned how he did that trick.'

William Stearn:
INSPIRATION FOR GARDENERS

❧ William Stearn had a brilliant mind, but without the traditional opportunities for formal education, he educated himself. He read, absorbed, reasoned, questioned and analysed, mixed with intellectual fellow enthusiasts, went to evening classes, and used all of the facilities available to him. It is perfectly possible to educate yourself as a gardener or botanist – there are many people who have done so, and this should be an inspiration to us all.

❧ Stearn wrote his first major monograph on the genus *Epimedium* and pubished a revised and greatly enlarged edition in 2002. His first mongraph published in 1938 included 21 species; the revised monograph included 54. These low-growing woodland plants are

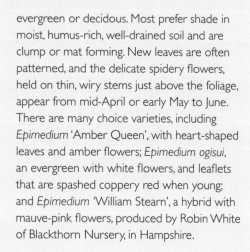

evergreen or decidous. Most prefer shade in moist, humus-rich, well-drained soil and are clump or mat forming. New leaves are often patterned, and the delicate spidery flowers, held on thin, wiry stems just above the foliage, appear from mid-April or early May to June. There are many choice varieties, including *Epimedium* 'Amber Queen', with heart-shaped leaves and amber flowers; *Epimedium ogisui*, an evergreen with white flowers, and leaflets that are spashed coppery red when young; and *Epimedium* 'William Stearn', a hybrid with mauve-pink flowers, produced by Robin White of Blackthorn Nursery, in Hampshire.

❧ Stearn became inspired by epimediums when browsing through plant hunter Reginald Farrer's *The English Rock Garden*. Farrer, who was renowned for his descriptive prose, characterised the blooms as 'Columbines of waxy texture, and in any colour, from white, through gold, rose and violet'.

❧ Stearn's monograph encompassed 'other Berberidaceae including the genus *Podophyllum*'. The latter enjoys the same 'woodsy' soils as *Epimedium*. One selection that has gained considerable popularity is the hybrid *Podophyllum* 'Spotty Dotty'. This has large, lobed parasol-like leaves, which are chartreuse green, marked with chocolate-brown spots. Under these hang large beetroot-coloured flowers, followed by plump red fruit. The fruits, though attractive, are poisonous and should not be eaten.

LEFT Stearn was a globally respected authority on the *Epimedium* genus. Try growing the lovely, hardy *Epimedium* 'Amber Queen' (barrenwort).

ABOVE Stearn explained in accessible terms how plants got their Latin names. *Rosa farreri* f. *persetosa* (Farrer's threepenny-bit rose) honours plant collector Reginald Farrer.

LEFT The colouring of *Mimulus guttatus* (monkey flower) is described in its Latin name; the translation of *guttatus* is 'with spots'.

Richard Evans Schultes

DATE	1915–2001
ORIGIN	UNITED STATES OF AMERICA
MAJOR ACHIEVEMENT	*PLANTS OF THE GODS*

Professor Richard Evans Schultes, considered by many to be the 'Father of Ethnobotany', spent much of his life in Amazonia, wrote the rainforest herbal *The Healing Forest*, and was an authority on hallucinogenic plants. His encyclopaedic knowledge of his subject led him to rebuke his Harvard colleague Timothy Leary for misspelling botanical names, and as a serious researcher and scientist, he showed little patience with those who indulged in mind-altering drugs. When William Burroughs described a trip as a 'mind-opening metaphysical experience', Schultes responded: 'That's funny, Bill, all I saw was colours'.

Lophophora williamsii,
dumpling cactus

Frrom the age of six, Richard Evans Schultes became passionate about plants, travel and South America, after his parents read him the travelogue of Richard Spruce, a nineteenth-century plant hunter in the Amazon and Andes while he was ill in bed with stomach problems for a considerable length of time.

On leaving school in 1933, he won a full scholarship to study medicine at Harvard but he soon swapped courses to pursue his first love – plants. Professor of Botany at the time, Oakes Ames, encouraged him to study ethnobotany, a field that Ames described as having 'tremendously important human implications'. As an undergraduate, Schultes read *Mescal: The Divine Plant and Its Psychological Effects* by German psychiatrist Heinrich Kluver, about a small grey-blue, hallucinogenic cactus *Lophophora williamsii* (also known as the dumpling cactus, or peyote – pictured on p194) and he set off for Oklahoma to study its use by Kiowa Indians. For his post-graduate thesis, he headed south to Oaxaca in Mexico, studying the native peoples' use of sacred plants such as the ololiuqui (*Rivea corymbosa*) and the hallucinogenic mushroom teonánacatl (*Panaeolus campanulatus* var. *sphinctrinus* and *Psilocybe cubensis*); later, compounds extracted from the fungi led to the manufacture of the first beta-blockers.

After completing his doctorate in 1941, Schultes received a grant from the National Academy of Sciences to study curare, a rainforest drug used in surgery, whose properties as a muscle relaxant avoided the need for deep anaesthesia and was used as a treatment for tetanus. The exact composition of curare had been a mystery but by living with tribes in the Colombian Amazon, Schultes identified over 70 plant species involved in its production; several of them are now known to contain active alkaloids, notably *Chondrodendron tomentosum*. Schultes also discovered that the combination of species varied according to the effect being sought, and that what was once thought to be a primitive cocktail of poisons was in fact a complex indigenous pharmacopeia.

FOLLOWING IN SPRUCE'S FOOTSTEPS

Schultes travelled the Amazon's tributaries in an aluminium canoe, sometimes alone, and at other times accompanied by a native guide. On one journey up the river he hit some rapids and his canoe overturned; he scrambled back inside but lost his guide and all his equipment. Weakened by malaria and beriberi he paddled for ten days before finding help at a remote outpost. All he carried on his trips was a change of clothes, notebooks and pencils, a pith helmet, a hammock, a machete and a small medicine case containing vitamins, antibiotics and morphine – in case he broke a limb and had to be transported for days before receiving proper treatment. He went native with his diet apart from regular supplies of instant coffee and baked beans, preferring to rely on food offered by his indigenous friends. This included ground manioc roots, fish, wild game, insect grubs, fruit and chicha, a drink made from fruit that is chewed and fermented in spittle. Schultes also carried a pair of secateurs, harvesting up to 90 specimens a day, and a camera

and film to photograph the tribes he lived among. Some of the images were published in 1988 in his book *Where the Gods Reign*. He did not believe the indigenous people were hostile, stating instead: 'All that is required to bring out their gentlemanliness is reciprocal gentlemanliness.'

His friendships with local shamans gave him a great understanding of the rainforests as their larder, laboratory and medicine cabinet. Many plants they showed him were unknown to science, and over 3,000 were named after him. *Hiraea schultesii* is used to cure conjunctivitis, *Piper schultesii* treats tubercular coughs, *Schultesianthus* is a genus in the potato family, and *Pourouma schultesii* treats ulcers and wounds. When taxonomists ran out of ways to use his name they resorted to using his initials, naming a new genus in the African violet family, *Resia*. However, Schultes wore the honour lightly, and the only one he ever spoke about was the firefly cockroach (*Schultesia lampyridiformis*) and he carried a picture of it around in his wallet.

Schultes identified *Strychnos toxifera* (curare) as a source of a muscle relaxant, so this species contains both potent arrow poison and medicine.

On hearing of the Japanese attack on Pearl Harbor, Schultes headed off through the rainforest for Bogotá to enlist, but on arrival was told he was needed elsewhere. The Japanese occupation of Malaya had deprived the Allies of their main source of rubber, originally an Amazon rainforest tree. With an intimate knowledge of the forest and its people he was perfectly placed to investigate alternative sources. He set up a rubber plantation that was managed and tapped by Indians and collected thousands of specimens of *Hevea*, selecting them for disease resistance and yield, and remaining in the Amazon until 1953, apart from a few visits home.

THE HARVARD PROFESSOR

Schultes's explorations yielded over 24,000 plant specimens, and of these, 2,000 had practical uses, ranging from medicines to clothing and contraceptives. He was one of the first campaigners to fight for the rainforest and its people, racing to record as much of their knowledge as possible before 'civilisation' arrived.

Returning to Harvard in 1953, Schultes became Curator of the Oakes Ames Orchid Herbarium, then the Curator of Economic Botany, later becoming Executive Director of the Harvard Botanic Museum and Edward C. Jeffrey Professor of Biology, posts he held until he retired in 1985. He also received the Gold Medal from the World Wildlife Fund and another from the Linnean Society. The Colombian government awarded him its highest honour and named a 22.2-million-acre area of protected rainforest 'Sector Schultes'.

Richard Evans Schultes often described himself as 'just a jungle botanist', yet he could read, write or speak ten languages, including two Amazonian languages – Witoto and Makuna – and when lecturing at Harvard, he taught his students to shoot a dart into a target on the other side of the lecture hall, using a blowpipe.

Richard Evans Schultes:
INSPIRATION FOR GARDENERS

❦ Before Richard Evans Schultes confirmed that ololiuqui was *Rivea corymbosa*, nineteenth-century botanists had believed it to be *Ipomoea sidaefolia* (now known as *Turbina corymbosa*), the genus that includes morning glory. Several are grown in gardens as climbers that can be grown as perennials in warmer climates, or half-hardy annuals in a cooler location. Morning glory behaves like its relative bindweed when hot summers or warm climates remind them of their native habitat.

❦ Speed of growth makes morning glory useful for new gardens, providing temporary screening or covering unsightly buildings. If the micro-climate allows for overwintering outdoors, the cumulative effect can be spectacular. *Ipomoea indica* was planted at least 20 years ago by the side of a building in the courtyard at Lambeth Palace in London. Protected from frost, it has now reached the top of the wall and is around 20m (65 1/2ft) tall.

❦ Best planted where it is backlit by the sun, morning glory needs a sheltered sunny spot and room to roam. Soak seeds for 24 hours before planting, then sow at 20°C (68°F) in small pots or modules, to avoid root disturbance when transplanting. Grow them on in cooler conditions, 'hardening off' when

TOP LEFT In 1941, Schultes documented the use of *Ipomoea* by native Americans in Mexico, which he traced back to Aztec times. *Ipomoea purpurea* 'Grandpa Ott' is a vivid cultivar for the garden.

LEFT Vibrant *Aphelandra aurantiaca*, used by the South American Ticuna to treat deafness, was in *The Healing Forest* (1990) by Schultes and Robert Raffauf.

temperatures rise – they are particularly sensitive to cold when young and show their disapproval with checked growth and pale yellow or white leaves; they often never recover.

✤ Schultes charted the native-American use of *Ipomoea* from Aztec times to the present day. All plants from the genus are colourful for the summer garden; a lovely option is dark-purple-flowered *Ipomoea purpurea* 'Grandpa Ott'. This was given to Kent and Diane Whealy (née Ott) in 1972 by her grandfather, Baptist John Ott. His parents were Bavarian immigrants who sailed to America in 1867, bringing seeds of morning glory that they had grown in their garden in Germany. It grew over the porch on their 40-acre farm near St Lucas, Iowa, and on what proved to be their final visit, Grandpa Ott gave Diane some seeds. These (and those of a heritage tomato) were the catalyst for her interest in heirloom varieties, and the beginning of Seed Savers Exchange in the USA.

TOP Schultes studied the use and effects of *Psilocybe mexicana* (hallucinogenic mushroom). Related species can appear with other fungi on damp lawns.

ABOVE *Strychnos usambarensis* has the same use as the *Strychnos* arrow poison plants Schultes studied in the Amazon, but this species is a native of Africa.

Mikinori Ogisu

DATE	BORN 1951
ORIGIN	JAPAN
MAJOR ACHIEVEMENT	*TRADITIONAL FLORICULTURE IN JAPAN*

Mikinori Ogisu, a botanist, horticulturist and plant explorer, is renowned
for his introductions of new or long-lost plants. He has travelled extensively
in China, covering hundreds of thousands of kilometres, particularly in
Sichuan, and has visited plant-rich Mount Emei many times. Taxonomic
botanists have benefited greatly from his work and introductions of new
plants; *Epimedium* and *Mahonia*, in particular, have enriched our gardens.
He is also an authority on Japanese horticulture and ornamental plants,
and has achieved the status of 'national treasure' in his home country,
in recognition of his work.

Iris laevigata,
Japanese iris

Mikinori Ogisu was born in Owari, Aichi Prefecture, Honshu Island. He revealed a passion for plants when he was only 7; received private tuition from experts in the propagation, production and cultivation of traditional Japanese ornamental plants at 10; and by 15 he had set his mind on a career in horticulture and landscape design. His guru, Kazuo Anami, who taught natural sciences at his high school, had seen most of the 5,500 species of Japanese native plants, was a talented botanical illustrator and authored several books on the Japanese flora.

On leaving school, Ogisu broadened his knowledge and experience with a year at the Keisei Rose Research Institute, studying genetics and plant breeding. At the age of 20, he went to Europe to study at the Kalmthout Arboretum in Belgium, and then the Royal Botanic Gardens, Kew, and the RHS garden at Wisley, before enrolling on a postgraduate course in the department of biology at Sichuan University. There he became an honorary research scholar, and like Professor Fang, his tutor, developed an affection for Mount Emei and its flora. Ogisu has made well over 30 visits to the mountain, maintaining a checklist of its flora, confirming early records and adding new finds; his knowledge of the flora of Sichuan, and particularly of Mount Emei, is unrivalled.

MASTER OF JAPANESE HORTICULTURE

In 1977 the publication of *History and Principles of Traditional Floriculture in Japan*, an account of 33 groups of traditional ornamental plants from the Edo period, secured Ogisu's reputation as one of the most remarkable botanists in Japan. In 1978, the late Seizo Kashioka, founder of the Kansai Tech Corporation, wanted to make a contribution to Japan's horticultural heritage. Inspired by Ogisu's passion for traditional Japanese ornamentals, and aware of the threat to their survival, he established the Japanese Iris Garden at Yamasaki Shiso-gun, west of Osaka. Under Oguisu's guidance, the garden was filled with 1,000 cultivars and hybrids of *Iris ensata* var. *spontanea*, and later expanded to include *Iris laevigata* (pictured on p200), hydrangeas, evergreen azaleas and *Phlox subulata*. In 1986 a research institute was established on the site and collections expanded to include *Hosta*, *Asarum*, *Nandina*, *Epimedium* and many other genera, a collection of well over 3,000 different species and cultivars. Ogisu was retained as special advisor.

Since 1980, Ogisu has made annual visits to China (sometimes five or six a year), travelling many hundreds of thousands of kilometres, often to remote, little botanised areas, discovering new species and rediscovering rarities. In the early 1990s, Professor William Stearn (pp188–193), pondering an update of his 1938 monograph on *Epimedium*, was eager to discuss the taxonomic status and distribution of Chinese species with him. When Ogisu, who had already discovered and introduced several undescribed species, presented Stearn with several beautifully described specimens and an album of colour photographs displaying dissections of the flowers of a dozen or more taxa, the decision became easy. Stearn revised his early work,

described the new species, and new plants were introduced into cultivation, including *Epimedium fangii*, *Epimedium franchetii* and *Epimedium leptorrhizum*. Stearn wrote of him: 'no man has made a greater contribution to the knowledge of the genus *Epimedium*, notably in the province in Sichuan by introducing live plants for study in cultivation and collecting herbarium specimens, and photographing plants in their habitats, together with their floral parts than Mikinori Ogisu.'

NOTABLE REDISCOVERIES

Ogisu's wide-ranging interests led to a host of introductions and reintroductions into cultivation. In 1983 Graham Thomas challenged him to rediscover the legendary wild rose, *Rosa chinensis* var. *spontanea*, unseen by Western botanists since plant explorer Joseph Rock collected herbarium specimens early in the twentieth century. In autumn that year, Ogisu was driving his jeep at dusk along treacherous winding roads in a mountainous area of southwest Sichuan, when out of the corner of his eye he spotted some late-flowering roses. Despite being tired, despondent and unwell, he parked his jeep precariously on the verge and climbed 200m (656ft) uphill to investigate. His instinct was rewarded – it was the long-lost plant. In March 2005 he discovered a daphne, again from his car. A broken flowering stem with yellow flowers lying on the roadside alerted him to plants on the slopes above. Convinced it was a new species he took the specimen back to Japan, checked existing literature and on consultation with Li Zhenyu of the Beijing Institute of Botany at the Chinese Academy of Sciences, it was named *Daphne ogisui*.

Epimedium plants including: E. hybridum, E. versicolor and E. sulphureum. Ogisu greatly increased the choice for gardeners within the genus.

In 1989, Ogisu accepted yet another challenge: to find Tibetan hellebore (*Helleborus thibetanus*). First discovered by nineteenth-century French missionary and plant explorer, Père David, it was little known to anyone except local people who used its roots in herbal medicine. Using David's writings as a guide, Ogisu arrived in Moupin (now Baoxing) in west China. On showing a dried specimen to farmers, he was met with a blank, but when he described 'a big pinkish flower, growing straight out of the snow', several people immediately responded and led him to rocky ground where they were growing, quite possibly in the same site David found 120 years before.

In 1994 Ogisu was awarded the K. Matsushita Expo '90 prize for his work in botany and horticulture. In 1995, he received a Veitch Memorial Medal from the RHS, and the Yoshikawa Eiji Cultural prize recognising his botanical achievements in China. He also wrote about his explorations in a 2002 book published in Japan called *In Search of Long Lost Plants*.

Ogisu remains respectful of people, plants and nature, and, in turn, highly respected by his peers. An ardent defender of Japanese horticultural traditions and garden plants, with a keen eye for detail, he more than lives up to his sobriquet, the 'Green Samurai'.

Mikinori Ogisu:
INSPIRATION FOR GARDENERS

❖ Plants named after Mikinori Ogisu, his family and associates include several garden-worthy members of the *Epimedium* genus, among them *E. mikinorii*, with dainty rose-purple and white flowers; *E. ogisui*, boasting large white flowers; and *E. leptorrhizum* 'Mariko', with pale magenta-and-white flowers, which was named by Ogisu after his wife. There is also *E. sternii*, with caramel-and-pink flowers, and *Epimedium* 'William Stearn', with dark red and scarlet flowers with yellow spurs, named after Ogisu's great friend and collaborator. In Sichuan, Ogisu discovered what is now *Cotoneaster ogisui*, with white flowers followed by large red fruits; the branches of the small tree bend under the weight of the display and are followed by attractive autumn colour. There is also *Prunus incisa* 'Mikinori' (cherry 'Mikinori'), named after him by Harry van de Laar in 1989. This is an early flowering shrub, covered in spring with semi-double flowers, white to pale pink, with leaves that turn to shades of yellow, purple and red in autumn. And from southern Sichuan there is the attractive and distinctive honeysuckle *Lonicera subaequalis* 'Ogisu', with yellow flowers in tubular clusters, surrounded by a hooded green cup.

❖ Ogisu has also focused on mahonia species, adding *Mahonia ogisui* (boasting large, glossy leaflets on long leaves, and stout spikes of yellow flowers), *M. sheridaniana* (bearing long, finger-like spikes of bold yellow flowers) and a hybrid found in the wild and cultivation, *M.* x *emeiensis* (which varies in leaf shape and the density of waxy coating on the leaf underside).

BELOW *Epimedium ogisui*, a perennial with bright white flowers, was named by botanist William Stearn to honour Ogisu's contribution to the genus.

BELOW *Epimedium mikinorii* was collected by Ogisu in Hubei, China, and then named for him. It is evergreen except in severe winters.

❖ *Bergenia emeiensis* is found on limestone cliffs, often under overhangs where water seeps through the rock. First collected in 1935, it was introduced to cultivation in the West by Ogisu, who collected it on Mount Emei in 1982. When first introduced it was regarded as tender and grown in pots under cover, to protect and encourage the flowers and to keep its attractive dark green leaves undamaged through winter, but the plant itself is completely hardy. This elegant species can be grown outdoors in a protected position.

TOP *Mahonia bodinieri*, introduced to cultivation by Ogisu in 1997, can be found growing in the Queen Elizabeth II Temperate House at the Savill Gardens in Windsor, where it is used to add winter colour.

RIGHT Ogisu found *Bergenia emeiensis* growing on cliffs in China. Although hardy, it prefers a shady, protected spot outdoors and humus-rich soil.

Bleddyn & Sue Wynn-Jones

DATE	BLEDDYN BORN 1948; SUE BORN 1952
ORIGIN	WALES
MAJOR ACHIEVEMENT	PLANT COLLECTORS

A twist of fate led Bleddyn and Sue Wynn-Jones into horticulture, but now their nursery, Crûg Farm Plants in North Wales, with its reputation for rarities (including *Holboellia latifolia*, pictured below), has become the focal point for anyone with a passion for desirable plants. Their plant-collecting exploits around the world have created a catalogue full of new introductions and discoveries, particularly those that tolerate cool temperate climates. Their work is benefitting not just gardeners, who like to collect and grow them at home, but government agencies, taxonomists and conservation organisations, too.

Holboellia latifolia

Bleddyn and Sue Wynn-Jones changed from beef farming to plants with the impending arrival of BSE in 1990. As Bleddyn had been a keen vegetable grower and they loved travelling, they began producing hardy geraniums, exporting *Tropaeolum tuberosum* var. *lineomaculatum* 'Ken Aslet' and *Tropaeolum speciosum* to the Dutch, then turned to plant hunting after deciding to specialise in selling unusual plants. Most of the locations where they collect have a climate similar to northern Europe, where the greatest killer is winter wet.

Their first venture, a three-month trip to Taiwan, in 1992, was aided by a Taiwanese friend whose husband worked for the National Park Service. He organised permissions to collect in their target areas, including Taroko National Park, in the eastern part of the country where mountains rise above 3,000m (9,842ft). Before leaving on their trip the Wynn-Joneses visited Treborth Botanic Gardens, at Bangor University, to learn techniques for collecting, recording and taking herbarium specimens; Sue also made 200 cloth bags to take with them (they now use paper and plastic zip-lock bags). Most of the 1,500 seeds and plants they collected grew successfully, including a new species, *Actaea taiwanensis*, with spikes of scented white flowers. The area where they collected later disappeared in an earthquake; it is not unusual for whole mountainsides to drop away, and with that, their flora is lost forever.

VETERAN COLLECTORS

The Wynn-Joneses have visited over 30 countries, among them Laos, Guatemala and Colombia and many others that have been poorly botanised. The couple share their knowledge with scientists and government agencies in these countries, who cannot afford to send out botanists of their own, and as part of their policy of giving back to the community, they always employ local guides.

Their collaboration with the Vietnam Environment Agency has been particularly rewarding. In 2003 Bleddyn discovered a botanical 'hot spot' but had little time to collect. Returning three years later, the couple found an epiphytic lily, growing in moss on vertical trunks and horizontal branches of large forest trees, well above the ground, at altitudes of 1,900 to 2,000m (6,233–6,561ft). Many thought it to be long-lost *Lilium arboricola*, introduced by Francis Kingdon-Ward (see pp164–169) in 1953 but lost to cultivation. However, Bleddyn believed it was something different, and he was right; it was a new species of lily (later named *Lilium eupetes*), with maroon-purple flowers. They also discovered what is believed to be a new genus, a tree with pink flowers in the family Hamamelidaceae, which has been incorrectly named *Disanthus ovatifolius* by an overzealous Russian collector. Conservation has become a massive part of the Wynn-Joneses' work, since it is a race against time to save plants like these – the Vietnamese government is encouraging people to settle in these habitats and they are rapidly being cleared of understory plants to grow black cardamom. Plants are being lost before being described, so Bleddyn and one of their employees now make more freqent, shorter trips to Vietnam.

Days are long, and climbing steep mountains or travelling through dense jungle is physically demanding. At the end of each day, collections are recorded, photographs are captioned, often in tents battered by unforgiving weather. Plants are pressed and given an accession number, then as much detail as possible is written in a book and keyed into a laptop – soil type, altitude, location in sun or shade. GPS positioning provides exact locations and a mobile phone, laptop and Google Earth combine to reveal routes in remote locations. The Wynn-Joneses also use their phones to show people pictures of the plants they are seeking.

The Wynn-Joneses have made over 18,000 different collections to date. They began by looking mainly for herbaceous plants for shade, then woody plants, learning much from American plantsman and gardener Dan Hinkley when they collected together in the early days. The Wynn-Joneses taught Hinkley about herbaceous plants, and Hinkley taught them about woody plants. They then focused on the Araliaceae family, and genera like *Schefflera*, in a quest for plants with impact, discovering architectural *Schefflera taiwaniana* in the mountains, with up to 11 oval to oblong long, leafleted leaves and elegantly long purple leaf stalks, and *Fatsia polycarpa* in woodlands, with large, many-lobed, matt green leaves. Both look exotic and tender, but are actually tough plants.

THE CONTINUING QUEST

Since 2017, the couple have explored for plants in Georgia, looking for *Ruscus colchicus*, a highly suitable species for British horticulture. Working alongside botanists and taxonomists from the Moscow State University and the Batumi Botanical Garden in Georgia has allowed the Wynn-Joneses to collect previously unseen forms and varieties. Some Georgians are wild harvesting this plant for the cut-flower trade and flying them out to Moscow, a practice that is in the process of being regulated, since *R. colchicus* is rare in Georgia.

Bleddyn and Sue Wynn-Jones collected *Echeveria bicolor* in the Colombian Andes at 3,600m (11,811ft) above sea level, at El Cocuy in 2015.

The Wynn-Joneses have been threatened by Maoist guerrillas in the Himalayas, spent the night in an old paddy field infested by thousands of leeches (only dissuaded from streaming into the tent by an application of insect repellent to the vents), have eaten anteater and have endured many of the hardships experienced by the early plant explorers. The latest challenge is legislation – the Convention on Biodiversity (CBD) has not only increased the amount of bureaucracy, but many countries also try and use it to generate 'additional' income.

Nevertheless, the couple find their travels extremely rewarding, and both enjoy experiencing different cultures. But it is the plants that matter above all. As Bleddyn has explained: 'There is no point in bringing back plants that are already in cultivation, unless they are new forms, but the pleasure is returning with surprises and bringing back plants with impact, that are of benefit to science, conservation and gardeners.'

Bleddyn & Sue Wynn-Jones:
INSPIRATION FOR GARDENERS

✤ One of Bleddyn Wynn-Jones's favourite finds is *Pachysandra axillaris* 'Crûg's Cover', collected in Sichuan in 2000 with Dan Hinkley after a trip with the Alaskan chapter of the North American Rock Garden Society. It was found on the lower slopes of Longzhoushan, in saturated ground under a small colony of *Hydrangea chinensis.* Very different from the well-known *P. terminalis,* it has scented flowers, like *Sarcococca,* on short, dense, upright spikes,

RIGHT The Wynn-Joneses collected *Acer amoenum* on Mount Matsuo in western Honshu, Japan, in the autumn of 2005. It is easy to grow and can be 4m (13ft) tall.

BELOW *Pachysandra axillaris* 'Crûg's Cover', collected by Bleddyn, has proved to be very popular with gardeners. Resembling a *Sarcococca,* it has fragrant flowers, attractive glossy leaves and pink-red fruit.

and polished, undulating leathery leaves on short flexible, suckering stems. It performs very well as vigorous ground cover in moisture-retentive, fertile soil, in light to deep shade.

✤ French botanist Patrick Blanc fell in love with *Schefflera taiwaniana* when he visited Crûg Farm Plants. Arriving in the dark, he opened the shutters the following morning and could not believe what he saw. He was so excited about its structure he ran straight out into the garden to take a closer look. This small and elegant evergreen tree or large shrub, reaching 3 to 4m (9³/4–13ft) tall, was one of the Wynn-Joneses' first wild collections, from the high mountain forests of central Taiwan. Each long purple leaf stalk bears 7 to 11 oval to oblong leaflets. Green-tinted flowers are followed by purple fruit in winter.

✤ One of Sue's favourites is *Dahlia excelsa* 'Penelope Sky'. When she suggests to visitors they buy dahlias they sometimes turn up their noses, but when they see this plant they cannot believe its lovely foliage, beautiful flowers, and the height and structure of its bamboo-like stems. It was discovered on the moist mountains to the east of Oaxaca, southern Mexico, at around 2,500m (8,202ft) and named for their youngest granddaughter. The upper parts bear bronzy pinnate foliage and large lilac-purple flowers from July to the first frost. It is ideal for a warm, sunny spot in moist, fertile, well-drained soil, protected from severe frost.

TOP RIGHT *Heloniopsis umbellata* is a rare clump-forming woodland perennial, introduced by the Wynn-Joneses and now prized in gardens.

RIGHT *Dahlia excelsa* (giant dahlia). Sue often recommends 'Penelope Sky', a cultivar of this species; it grows to 3m (9³/4ft) and it flowers from midsummer until the first frosts.

Kingsley Dixon

DATE	BORN 1954
ORIGIN	AUSTRALIA
MAJOR ACHIEVEMENT	CONSERVATION AND LANDSCAPE RESTORATION

Professor Kingsley Dixon's passion for plants was ignited as a child when he would disappear for hours during school holidays exploring the bush around his home in the suburbs of Perth, Western Australia – much to his parents' consternation. It was further fired up when he began working in a native plant nursery at the age of 13. Dixon later studied botany at university, thriving in the academic environment and, decades on, he continues to apply his scientific knowledge and practical love of gardening to plant conservation and landscape restoration while sharing his boundless enthusiasm and knowledge of native plants with others.

Pterostylis banksii,
greenhood orchid

Professor Kingsley Dixon was raised in Morley, Western Australia, spending much of his childhood exploring bushland around his home; but what was once a botanical treasure trove is now an ocean of roofs to the horizon, its flora, fauna and complex ecosystems crushed by bulldozers and replaced by suburbia. These memories became the foundation of his commitment to conservation.

Dixon came from a family of keen gardeners; his father collected water lilies, his mother was always eager to help and his paternal grandmother became his gardening inspiration. Aged 6, he was given his own area within the shade house – a glasshouse being well beyond the family's meagre budget.

His eureka moment came in 1966, on coming face to face with Sir Joseph Banks (see pp68–73) surrounded by an impressive array of native nuts and seeds on Australia's new $5 note. Impressed by the fact that it was possible to love native plants yet be respected enough to feature on currency, he subconsciously began his botanical career.

Aged 13, Dixon spotted the sign for Wyemando Native Plant Nursery through the car window on his way to his holiday job collecting bottles at the tip where his father was a bulldozer driver. He demanded to be dropped off, rang the bell and met sisters Nan and Sue Harper who went on to teach him Latin names, their meaning and how to cultivate Australian native species.

A BOTANICAL EDUCATION

Dixon began growing orchids at the age of 12, and his parents later joined him. They amassed a large collection of Cymbidiums, while young Kingsley relished the challenge of growing difficult genera. Dixon mounted displays of plants and orchids at school and, when responding to questions about his future, said: 'I would like to be a botanist'. It was suggested he enrol on an Agricultural course at the University of Western Australia but after passing a building with 'Botany' emblazoned on the front, he enrolled for a BSc in Botany instead.

He thrived at university, and during his final year he corresponded with Dr Jack Warcup, an authority on native orchids, who sent him the mycorrhiza of a greenhood orchid (pictured on p212). Entranced, he learned techniques of symbiotic culture which is necessary for the survival of both orchid and fungi. During his honours degree, he devised a method of labelling vinegar flies with a heavy isotope of nitrogen, proving Darwin's theory that sundews use carnivory to obtain nitrogen. A PhD followed on The Ecophysiology of Geophytes, allowing him to study his favourite ground orchids of Western Australia; boasting around 450 species, it is one of the richest in the world.

Six months' unemployment followed until from sheer desperation, he convinced the plant pathologist at the University of Western Australia that he could work on Phytophthora fungal disease in *Banksia*.

Later, a WWF grant funded his search for the famed Western Australian underground orchid, *Rhizanthella gardneri*, which he had seen in orchid books and

as a pickled specimen in the State Herbarium as a child. Other members of the Western Australian Native Orchid Study and Conservation Group joined the forays. In 1982–3 they found 180 specimens, the halcyon days.

In 1981 he applied for a post as Research Assistant to the Director at Kings Park and Botanic Garden, Perth, and was devastated not to get the job. Months later they contacted him again, asking him to be Display Botanist and he gleefully accepted, despite being based in a tumble-down tin shed 'laboratory' with broken equipment. It became his first opportunity to share his enthusiasm for botany with the public. His first research grant in 1983, to study *Leucopogon obtectus*, the hidden beard heath – a rare shrub only found north of Perth – further energised his passion for research and native plant conservation.

Dixon became adept at applying for grants to fund research and ecological restoration projects. Ten years later at Kings Park there were 15 postgraduates and researchers, with $160,000 sponsorship funding a purpose-built laboratory which became home to programmes on rare plant conservation, disease, orchids, germination ecology and physiology. Dixon's team were first to develop sophisticated tissue culture and cryostorage – freezing seed from threatened species in liquid nitrogen so that they survived almost indefinitely. 'Science-into-practice', primarily the restoration of habitats destroyed by mining, has become a hallmark of his work.

Cymbidium parishii has fragrant white flowers, with deep red markings on the lip.

SMOKE

Dixon had long been fascinated by mass germination and flowering after bush fires, and arguably his greatest achievement following a lead from South African researchers, with the help of colleagues from the University of Western Australia and Murdoch University, was the discovery of the critical role of smoke in germination in Australian plants. The analysis of over 4,000 chemicals led to the discovery of the critical molecule, karrikinolide, (after 'karrik', the Noongar word for 'smoke') and karrikinolides which, in 2004, became the first new group of plant growth hormones to be discovered in almost 30 years. He also found that plants in other parts of the world germinate after exposure to smoke, proving that this characteristic is not exclusive to plants in fire-prone regions. Thanks to Dixon's discovery, seeds of over a thousand different Australian native species can now be germinated and returned to the wild or gardens. He has overseen over 30 ecological restoration projects at over 18 mines in 15 different ecosystems from forests to savanna, encompassing thousands of hectares of once-degraded habitat.

The orchid *Caleana dixonii* has been named for him. In 2023 he was awarded the Order of Australia, for services to conservation, restoration science and teaching in the hope that his passion will inspire future generations to embrace a love of botany and the environment and experience the excitement of growing plants for themselves, as he first did as a child.

Kingsley Dixon:
INSPIRATION FOR GARDENERS

❖ Dixon planted his first *Sphaeropteris cooperi* (commonly known as *Cyathea cooperi*) aged 13 and his garden at Cypress Farm now has over 1,500 specimens of this Australian native tree fern. Slow growing, but well worth the wait, it is a wonderful single, thin-stemmed species, topped by a plume of elegant fronds. Grow in shade or part shade in frost-free conditions, keep the trunk damp using drip irrigation all year round unless growing in humid conditions; keep frost free in cooler climates.

BELOW *Sphaeropteris cooperi* is commonly grown throughout the tropics and subtropics for its lacy fronds, reaching up to 5m (16ft), and 'coin spotted' markings on the trunk, where fronds have been shed. Grow under protection in cooler climates.

❖ The *Chlorophytum* species has been one of Dixon's go-to plants since he first grew it when he was around 5 years old. This is an excellent houseplant for beginners. Easy to grow and propagate, it produces living baby plants at the tip of long stems, which are a constant fascination and ensure an endless supply of plants. Evergreen with several interesting cultivars, it needs frost-free conditions.

❖ *Anigozanthos manglesii* – the red and green kangaroo paw – is the plant Dixon most admired on his childhood bushland rambles, for its extraordinary colour combination, felt-like leaves and the fact it is bird pollinated. This is a staple in sand land restoration around Perth. It needs frost-free growing conditions; grow

outdoors in moist, well-drained humus-rich sandy soil, mulch in winter or grow under glass. (See also Georgiana Molloy, pp98–103.)

❖ A specimen of *Nymphaea* 'Conquerer' was planted in the first large, water lily pond in the Dixon family garden. It is fully hardy and has large flowers which appear reliably. Dixon's grandmother was a big fan of 'cow poo' gardening: water lilies were fertilised with golf-ball sized lumps of cow manure encased in clay pushed into the mud at the base of the pond. Gardeners can use more modern types of fertiliser, feeding annually in spring.

❖ *Portulacaria afra* is a slow growing, evergreen succulent often producing red flushed stems. This was the subject of Dixon's first attempt at bonsai at age 13. He grew it in a flat seed tray – his make-believe bonsai pot. It needs well-drained cactus compost or similar, in bright, indirect light. Water well from mid-spring and mid-autumn; keep almost dry at other times.

BELOW *Nymphaea* 'Conquerer' whose leaves reach 30cm (12in) across, needs plenty of space to grow and full sun to part shade in still water. It was introduced in 1910 by Joseph Bory Latour-Marliac, the genius behind Monet's water lilies.

BELOW *Portulacaria afra* is a widespread South African succulent that has been cultivated for centuries in Europe as a houseplant and is grown as a low succulent hedge in warmer regions of the USA. A good plant for beginners.

Bibliography

Sources consulted during research include books, scientific papers, websites and journals, primarily obtained through JSTOR, the Biodiversity Heritage Library and Hertfordshire libraries. Among the sources: the *Dictionary of National Biography*, the Royal Horticultural Society magazine, *The Garden, Curtis's Botanical Magazine, The Gardeners' Chronicle, The Plantsman, Garden History, Transactions of the Linnean Society, Transactions of the American Philosophical Society, Transactions and Proceedings of the Royal Society of New Zealand, Gardens Illustrated* and *Agricultural Bulletin of the Straits and Federated Malay States.*

Books include:

Barry, Bernice (2016) *Georgiana Molloy: The Mind That Shines*. Picador.

Bartram, William (1791) *Travels Through North and South Carolina, Georgia, East and West Florida*. James & Johnson.

Bellairs, Nona (1865) *Hardy Ferns: How I Collected and Cultivated Them*. Smith, Elder and Co.

Blunt, Wilfrid (2004) *Linnaeus, The Compleat Naturalist* (revised edition). Frances Lincoln.

Burbank, Luther (2004) *Some Interesting Failures: The Petunia with the Tobacco Habit and Others* (facsimile of 1914 edition). Athena University Press.

Dampier, William (2007) *A New Voyage Round the World* (facsimile of 1703 edition). 1500 Books.

Dandy, James (1958) *The Sloane Herbarium: An Annotated List of the* Horti Sicci *Composing it; with Biographical Details of the Principal Contributors*. British Museum.

Desmond, Ray (1999) *Sir Joseph Dalton Hooker: Traveller and Plant Collector*. ACC Art Books.

Desmond, Ray (2007) *The History of The Royal Botanic Gardens, Kew* (2nd edition). Royal Botanic Gardens.

Faulkner, Thomas (1813) *An Historical and Topographical Account of Fulham; Including the Hamlet of Hammersmith and Fulham*. J. Tilling.

Fuchs, Leonard (2016) *The New Herbal of 1543* (facsimile of 1543 edition). Taschen.

Gunther, Robert W. T. Ed. (1928) *Further Correspondence of John Ray*. Ray Society.

Hayavadana Rao, C. Ed. (1915) *The Indian Biographical Dictionary*. Pillar & Co.

Kingdon-Ward, Francis (1913) *The Land of the Blue Poppy: Travels of a Naturalist in Eastern Tibet*. Cambridge University Press.

Kingdon-Ward, Jean (1952) *My Hill So Strong*. Jonathan Cape.

Lazenby, Elizabeth Mary (1995) *The* Historia Plantarum Generalis *of John Ray: Book I – A Translation and Commentary*. Newcastle University.

Linnaeus, Carl (1737) *Genera Plantarum*. Conradum and Georg. Jac. Wishoff.

Linnaeus, Carl (1737) *Hortus Cliffortianus*. Amsteldami.

McLean, Brenda (2009) *George Forrest: Plant Hunter*. ACC Art Books.

Nelson, E. Charles (2014) *Shadow Among Splendours: Lady Charlotte Wheeler-Cuffe's Adventures Among the Flowers of Burma*. National Botanic Gardens of Ireland.

North, Marianne (1893) *Recollections of a Happy Life: Being the Autobiography of Marianne North*. Macmillan & Co.

Oliver, Francis. Ed. (1913) *Makers of British Botany*. Cambridge University Press.

Pavord, Anna (2005) *The Naming of Names: The Search for Order in the World of Garden Plants*. Bloomsbury.

Pringle, Peter (2008) *The Murder of Nikolai Vavilov*. Simon & Schuster.

Pulteney, Richard (1790) *Historical and Biographical Sketches of the Progress of Botany in England* (volume two). T. Cadell.

Raffauf, Robert and Schultes, Richard Evans (1990) *The Healing Forest: Medicinal and Toxic Plants of the Northwest Amazonia*. Dioscorides Press.

Smith, James Edward (1825) *The English Flora*. Longman, Hurst, Rees, Orme, Brown and Green.

Stearn, William T., (2002) *The Genus* Epimedium *and other Herbaceous Berberidaceae* (revised edition). Royal Botanic Gardens.

Stearn, William T. (1999) *John Lindley, 1799–1865: Bi-centenary Celebration Volume: Gardener – Botanist and Pioneer Orchidologist*. Antique Collectors' Club.

Stearn, William T. (1992) *Stearn's Dictionary of Plant Names for Gardeners*. Cassell Illustrated.

Turner, William (1548) *The Names of Herbes*. The English Dialect Society.

Turner, William (1995) *A New Herball: Parts I* (facsimile of 1551 edition). Cambridge University Press.

Wilson, Ernest H., (1917) *Aristocrats of the Garden*. Doubleday, Page & Company.

Index

Image credits

5, 166 © Lyubov Tolstova | Shutterstock. 8, 16 (bottom), 17, 26 (right), 56 (right), 68 (right), 74 (right), 86 (right) © Wellcome Collection. 12 © Ligak | Shutterstock. 13 (top) © VicW | Shutterstock. 13 (bottom) © nadtochiy | Shutterstock. 18 (top) © Sealstep | Shutterstock. 18 (bottom) © IanRedding | Shutterstock. 19 (top) © Niwat.koh | Shutterstock. 19 (bottom) © Olga Miltsova | Shutterstock. 20 © alex74 | Shutterstock. 23, 34, 74 (left), 106, 122 (left), 158 (right), 164 (right) © RHS | Lindley Library. 24 © blickwinkel | Alamy Stock Photo. 25 © Sophie Davies | Alamy Stock Photo. 30 (left) © Rob Whitworth Garden Photography | Alamy Stock Photo. 30 (right) © Ole Schoener | Shutterstock. 31 (top) © Gary K Smith | Alamy Stock Photo. 31 (bottom) © Elena Koromyslova | Shutterstock. 36 © FarOutFlora | Flickr Commons. 37 (top) © alybaba | Shutterstock. 37 (bottom), 42 © REDA &CO srl | Alamy Stock Photo. 43 (top) © Vladimir Gjorgiev | Shutterstock. 43 (bottom) © J Need | Shutterstock. 48 © haris M | Shutterstock. 49 (top) © Casliber | Creative Commons. 49 (bottom) © Geoff Derrin | Creative Commons. 52, 64 © The Natural History Museum | Alamy Stock Photo. 54 © Paul Asman and Jill Lenoble | Flickr Commons. 55 (top) © Graham Prentice | Shutterstock. 55 (bottom) © Ariel Bravy | Shutterstock. 60 (left) © Wonderful Nature | Shutterstock. 60 (right) © Anastasios71 | Shutterstock. 61 (top) © ncristian | Shutterstock. 61 (bottom) © A. Barra | Creative Commons. 66 © adaptice photography | Shutterstock. 67 (top) © John Richmond | Alamy Stock Photo. 67 (bottom) © RM Floral | Alamy Stock Photo. 72 © aniana | Shutterstock. 73 © haris M | Shutterstock. 78 (top) © Anna Gratys | Shutterstock. 78 (bottom) © Auscape International Pty Ltd | Alamy Stock Photo. 79 (top) © Martin Fowler | Shutterstock. 79 (bottom) © Eileen Kumpf | Shutterstock. 84 (left) © Gl0ck | Shutterstock. 84 (right) © RHS | Carol Sheppard. 85 (both) © Del Boy | Shutterstock. 90 © Sharon Day | Shutterstock. 91 (top) © Tamara Kulikova | Shutterstock. 91 (bottom) © Old Man Stocker | Shutterstock. 96 © skyfish | Shutterstock. 97, 169 (bottom) © Avalon/Photoshot License | Alamy Stock Photo. 102 © Universal Images Group North America LLC | DeAgostini | Alamy Stock Photo. 103 (top) © anastas_styles |

Shutterstock. 103 (bottom) © Jessika Knaupe | Shutterstock. 107 © De Agostini Picture Library | De Agostini | Getty images. 108 (left) © Kathy Clark | Shutterstock. 108 (right) © ZayacSK | Shutterstock. 109 (top) © Steffen Hauser | botanikfoto | Alamy Stock Photo. 109 (middle) © sarintra chimphoolsuk | Shutterstock. 109 (bottom) © Alpsdake | Creative Commons. 114 © RHS | Herbarium. 115 (top) © BENCHA STEWART | Shutterstock. 115 (bottom), 192, 204 (left), 210 (bottom) © RHS | Tim Sandall. 120 (left) © DGSHUT | Shutterstock. 120 (right) © J Need | Shutterstock. 121 (top) © RukiMedia | Shutterstock. 121 (bottom) © Steve Cordory | Shutterstock. 126 © ntdanai | Shutterstock. 127 (top) © Kathy deWitt | Alamy Stock Photo. 127 (bottom) © Julie Fryer pics | Alamy Stock Photo. 128 (right) © Paul Fearn | Alamy Stock Photo. 132 © RHS | Graham Titchmarsh. 133 (top) © ahau1969 | Shutterstock. 133 (bottom) © Alice Heart | Shutterstock. 136 (both) © Biodiversity Heritage Library | Flickr Commons. 138 © Jack Hong | Shutterstock. 139 (top) © sylviane decleir | Shutterstock. 139 (bottom) © Kenneth Keifer | Shutterstock. 144 (top), 211 (top) © Garden World Images Ltd | Alamy Stock Photo. 144 (bottom) © Peter Turner Photography | Shutterstock. 145 (top) © tamu1500 | Shutterstock. 145 (bottom) © Edgar Lee Espe | Shutterstock. 150 © Nature Photographers Ltd | Alamy Stock Photo. 151 © Rex May | Alamy Stock Photo. 152 (right) © RHS. 156 (left) © Layue | Shutterstock. 157 (top) © RHS | Rodney Lay. 157 (bottom) © HHelene | Shutterstock. 162 (top) © asdfawev | Flickr Commons. 162 (bottom) © Alexandra Giese | Shutterstock. 163 (top) © kukuruxa | Shutterstock. 163 (bottom) © Nate Allred | Shutterstock. 168 (left) © KirinX | Creative Commons. 168 (right) © Girija and Viru Viraraghavan. 169 (top) © travelview | Shutterstock. 174 © mpetersheim | Shutterstock. 175 (top) © Irina Borsuchenko | Shutterstock. 175 (bottom) © 271 EAK 7 Images. 180 © mizy | Shutterstock. 181 (top) © Alex Manders | Shutterstock. 181 (bottom) © Spring Images | Alamy Stock Photo. 212 (right) © Kingsley Dixon. 216 © David Clode | Unsplash. 217 (left) © Victoria Tucholka | Shutterstock. 217 (right) © Nooumaporn | Shutterstock. Portrait sketches on pp50, 92, 146, 182, 194 and 200 by Ian Durneen © The Bright Press. All other images are in the public domain.

Acknowledgments

I would like thank the following for their help during the preparation of this book: James Evans, Associate Publisher, and Lucy York, Senior Editor, at The Bright Press; Katriona Feinstein, Project Editor, for her endless patience and good humour; Lindsey Johns, for selecting such wonderful images; Julia Nurse, Collections Researcher at the Wellcome Library, for being so generous with her time and knowledge; Bernice Barry, expert on Georgiana Molloy, for her constant help and prompt replies to my many emails; Jessica Biggs, for her research and transcription; Jack Seddon and Hertfordshire Libraries, for facilitating library access; the RHS library: Fiona Davison, Crestina Forcina and Susan Robin, for providing books and texts; Dr Stephen Harris, Druce Curator (Herbaria), and Serena Marner, Herbarium Manager of Oxford University; Dr Isabelle Charmantier, Deputy Collections Manager and Librarian at the Linnean Society; Michael Marriott, Senior Rosarian; rose

breeders Calvin and Heather Horner; Deborah Jordan at Middleton Hall Trust; Dr Martin Gardner MBE, for his text on conifer conservation; Chris Lane of the RHS Woody Plant Committee; Tony Hall, Arboretum and Gardens Manager; Miranda Janatka, Collections Horticulturist at RBG Kew, and Leonie Paterson Archivist, at RBG Edinburgh; Oliver Tooley, for information on Frank Kingdon-Ward; Robert Vernon of Bluebell Nursery and Arboretum, for the story of *Liquidambar styraciflua* 'Lane Roberts'; Sorrel Everton and Mike Grant, for their help while researching Mikinori Ogisu; John Anderson, Keeper of the Gardens, and Harvey Stephens, Deputy Keeper of the Gardens, at Windsor Great Park; Hannah Newton, Dan Cocker and Laurence Bassett at Somethin' Else Productions; Martin and Nikki at The Local; Dan Wilson, Bill and Joy Parkin, Gill, Jessica, Henry and Chloe, for their help and support.